Louder Than Words

Louder Than Words

*The New Science of How
the Mind Makes Meaning*

Benjamin K. Bergen

BASIC BOOKS
A Member of the Perseus Books Group
New York

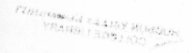

Published by Basic Books,
A Member of the Perseus Books Group

Books published by Basic Books are available at special discounts for bulk purchases in the United States by corporations, institutions, and other organizations. For more information, please contact the Special Markets Department at the Perseus Books Group, 2300 Chestnut Street, Suite 200, Philadelphia, PA 19103, or call (800) 810-4145, ext. 5000, or e-mail special.markets@perseusbooks.com.

Designed by Brent Wilcox

Library of Congress Cataloging-in-Publication Data
Bergen, Benjamin K.
    Louder than words : the new science of how the mind makes meaning /
Benjamin K. Bergen.
        p.   cm.
    Includes bibliographical references and index.
    ISBN 978-0-465-02829-0 (hardcover : alk. paper) —
ISBN 978-0-465-03333-1 (e-book)
    1. Meaning (Psychology)   I. Title.
    BF463.M4B47 2012
    153—dc23
                                                                    2012024985

10 9 8 7 6 5 4 3 2 1

Dedicated with love to Gerda.
Thank you for the tennis lessons.

CONTENTS

FOREWORD BY GEORGE LAKOFF

There is a revolution going on, a revolution in our understanding of what it is to be a human being. At stake is nothing less than the nature of the human mind.

For centuries, we in the West have thought of ourselves as rational animals whose mental capacities transcend our bodily nature. In this traditional view, our minds are abstract, logical, unemotionally rational, consciously accessible, and above all, able to directly fit and represent the world. Language has a special place in this view of what a human is—it is a privileged, logical symbol system internal to our minds that transparently expresses abstract concepts that are defined in terms of the external world itself.

I was brought up to think about the mind, language, and the world in this way. And I was there in the mid-1970s when the revolution started. Some philosophers, like Merleau-Ponty and Dewey, had already begun taking issue with the traditional view of the mind. They argued that—quite to the contrary of the traditional view—our bodies have absolutely everything to do with our minds. Our brains evolved to allow our bodies to function in the world, and it is that embodied engagement with the world, the physical, social, and intellectual world, that makes our concepts and language meaningful. And on the back of this insight, the Embodiment Revolution began.

It started with empirical research carried out mostly by analytical cognitive linguists who discovered general principles governing massive amounts of data. Certain computer scientists, experiment psychologists, and philosophers slowly began taking the embodiment of mind seriously by the 1980s. But by the mid-1990s, computational neural modelers and especially experimental psychologists picked up on the embodied cognition research—brilliant experimenters like Ray Gibbs, Larry Barsalou, Rolf Zwaan, Art Glenberg, Stephen Kosslyn, Martha Farah, Lera Boroditsky, Teenie Matlock, Daniel Casasanto, Friedemann Pulvermüller, John Bargh, Norbert Schwarz, and Benjamin Bergen himself. They have experimentally shown the reality of embodied cognition beyond a doubt. Thought is carried out in the brain by the same neural structures that govern vision, action, and emotion. Language is made meaningful via the sensory-motor and emotional systems, which define goals and imagine, recognize, and carry out actions. Now, at the beginning of the twenty-first century, the evidence is in. The ballgame is over. The mind is embodied.

The Embodiment Revolution has shown that our essential humanness, our ability to think and use language, is wholly a product of our physical bodies and brains. The way our mind works, from the nature of our thoughts to the way we understand meaning in language, is inextricably tied to our bodies—how we perceive and feel and act in the world. We're not cold-blooded thinking machines. Our physiology provides the concepts for our philosophy.

Every thought we have or can have, every goal we set, every decision or judgment we make, every idea we communicate makes use of the same embodied system we use to perceive, act, and feel. None of it is abstract in any way. Not moral systems. Not political ideologies. Not mathematics or scientific theories. And not language.

This is the first book to survey the compelling range of ingenious experimental evidence that shows definitively that the body characterizes the concepts used by what we call the mind. But the experiments do more than just confirm previous theory and description. They reveal that embodied cognition affects behavior. We act on the basis of how we think, and embodied thought changes how we perceive and how we act. As a society, we have to rethink what it fundamentally means to be human.

Louder Than Words is a stunningly beautiful synthesis of the new science of meaning. Benjamin Bergen offers a vivid, enthralling, and—remarkably—even funny introduction to the psychological experiments and brain research showing how your mind really works.

This book shows not only *that* actions speak louder than words, but *how*.

George Lakoff
Berkeley, CA
July 2012

CHAPTER 1

The Polar Bear's Nose

Polar bears have a taste for seal meat, and they like it fresh. So if you're a polar bear, you're going to have to figure out how to catch a seal. When hunting on land, the polar bear will often stalk its prey almost like a cat would, scooting along its belly to get right up close, and then pounce, claws first, jaws agape. The polar bear mostly blends in with its icy, snowy surroundings, so it's already at an advantage over the seal, which has a relatively poor sense of vision. But seals are quick. Sailors who encountered polar bears in the nineteenth century reported seeing polar bears do something quite clever to increase their chances of a hot meal.[1] According to these early reports, as the bear sneaks up on its prey, it sometimes covers its muzzle with its paw, which allows it go more or less undetected. Apparently, the polar bear hides its nose.

When I first read about this ingenious behavior, I found it fascinating.[2] Does the bear have the mental flexibility to envision what it looks like to others and the creativity to figure out how to conceal itself? Or is this nose covering just a trick that evolution has dropped into the polar bear's quiver of built-in behaviors—a freak behavior that happened to confer a survival advantage and was therefore selected for over the course of millennia?

Now, although there's doubtless a lot more to say about these charismatic megafauna, this is not a book about polar bears. It's a book about you and, more specifically, how you understand language. So consider, if you will, what you did when you opened this book and started reading the first paragraph. You cast your eyes over the letters that made up the words. You recognized familiar words like *bear* and *seal* and *hunting* and *snow*. That all seems pretty straightforward—it's the kind of thing a well-written piece of software or well-trained parrot could do. But then you started doing things that were a little deeper. Once you knew what the words were, you began to find meaning in them. You knew what type of animals and objects the nouns referred to and what types of actions and events the verbs described. But you didn't stop at the words. You made sense of the sentences they made up, sentences that I'm almost certain you had never encountered before (unless this not your first time reading this book). And the things the sentences described probably came to life—the bear scooting along its belly through the snow and the ingenious but awkward way it would have to hold its paw over its nose. Maybe you even went so far as to virtually "see" the arctic scene in your mind's eye.

And then—and here's the really remarkable part—you went way beyond that. You filled in details that were never explicitly mentioned. How do I know? You see, polar bears, as you surely surmised, hide their dark muzzles because the thick fur that covers their bodies, including their paws but not including their noses, is white. And they live surrounded by snow and ice, which for the most part is also white. But here's the thing. I actually never mentioned anything about color. If you look back at the first paragraph of this chapter, you'll see that the whiteness of the snow and of the bear and the blackness of its nose are completely implied. You colored in the picture. And it's a good thing you did, because without color, the story makes absolutely no sense at all. There's no other obvious reason for a polar bear to cover its nose.

How do you manage to do all this? How do you take scribbles on a page or for that matter the pops, buzzes, and hums of human speech and make them mean something to you? How do you know what the words and sentences mean and how do you fill in the gaps? How do you do what you're doing right now? That's the mystery of meaning. And <u>that</u> is in fact what this book is about.

The Meaning Makers

Making meaning might be one of the most important things we do. For starters, it's something we're doing almost constantly. We swim in a sea of words. Every day, we hear and read tens of thousands of them. And somehow, for the most part, we understand them. We understand who they refer to and what situations they describe. We make inferences about things that weren't even mentioned and prepare to respond appropriately. Constantly, tirelessly, automatically, we make meaning. What's perhaps most remarkable about it is that we hardly notice we're doing anything at all. There are deep, rapid, complex operations afoot under the surface of the skull, and yet all we experience is seamless understanding.

Meaning is not only constant; it's also critical. We use language to make sense of the world. We use it almost any time we interact with other people: to flirt, command, inform, beg, and form social bonds. A few words can change our minds, change our marital status, or change our religion. Words affect who we are. As a species, language is our most powerful and pervasive tool. With language, we can communicate what we think and who we are. Without language, we would be isolated. We would have no fiction, no history, and no science. To understand how meaning works, then, is to understand part of what it is to be human.

And not just human, but uniquely human. No other animal can do what we can with language. Of course, parts of human language have homologues in other animals. People talk fast, and sentences can be extremely complicated, but zebra finches sing tunes that rival our speed and complexity. Humans can drone on and on, but even a filibustering senator doesn't outlast humpback whales, whose songs can continue for hours. And although the human ability to combine words in new ways seems pretty unique, it's seen on a more limited scale in bees, who dance messages to each other that combine information about the orientation, quality, and distance of food sources. What's special about human language—what marks it as distinct from every other naturally occurring form of communication in the known universe—is that we can use it to convey pretty much any meaning that we want. A bee can waggle its abdomen until it falls off, but it will never communicate anything beyond what it's programmed to—it can't say that the weather's likely to clear up, that it had a decent night's sleep, or that it's looking forward to the weekend because it has a hot date with a hydrangea. Human language, in contrast to all other animal communication systems, is open-ended. We can talk about things that exist, like inarticulate presidential candidates and rail-thin models, or even things that don't, like Martian anthropologists or vegetarian zombies. And, for the most part, other people—at least people who speak our language and have normally functioning cognitive systems—are able to understand us. No other animal can do this. And because this level of meaning making is unique to our species, determining how it works brings us one step closer to knowing what distinguishes us from other animals.

There are other, more practical reasons to pursue the science of meaning. Imagine computer systems that truly understand you when you talk to them (Siri or Watson on steroids) or that can automatically translate from one human language to another. No reasonable *Star Trek*–worthy future would be complete without them. Understanding how meaning works can also help us improve the way we teach foreign languages. And it can lead to restorative therapies and technologies for people who have suffered brain damage that impairs their ability to understand or produce language meaningfully.

For all these reasons, language has held a privileged spot in science and philosophy throughout history. For centuries, philosophers have asked what is it that we humans have that our tongue-tied relatives don't; what cognitive capacities evolution has endowed us with that allow us to understand—and appreciate—sonnets and songs, exhortations and explanations, newspapers, and novels. And there are half a dozen academic disciplines dedicated to different aspects of language: from English and foreign languages to communications, semantics, psycholinguistics, cognitive linguistics, and neurolinguistics. Thanks to research in these fields, we now know a lot about the grammar of sentences, about how people articulate speech, and how to best teach a foreign language.

But for the most part, we've failed to answer the most important question of all. Language matters to us because it is a vehicle for meaning—it allows us to take the desires, intentions, and experiences in our heads and transmit a signal through space that makes those thoughts pop up in someone else's head. We don't study French in order to form perfectly grammatical French sentences; we learn it to communicate. We don't read fiction because the words look appealing on the page but because of the transporting flood of sights, sounds, places, and ideas that good writing evokes. And yet, almost no one, from lay people to linguists, really knows how meaning works.

That is, until recently. This is the age of cognitive science. Had we been born earlier, we might be exploring new continents. Born later, we might be gallivanting through the stars. But right now, at this time in our history, the vast, tantalizing expanse that begs to be discovered is the human mind. And some cognitive scientists, like me, have started to turn their attention to meaning. Over the past decade, a few key experimental advances have quickly elevated meaning to "hot topic" status in cognitive science. Using fine measures of reaction time, eye gaze, and hand movement, as well as brain imaging and other state-of-the-art tools, we've started to scrutinize humans in the act of communicating. We can now peer inside the mind and thereby put meaning in its rightful place at the center of the study of language and the mind. With these new tools, we've managed to catch a glimpse of meaning in action, and the result is revolutionary. The way meaning works is much richer, more complex, and more personal than we ever would have predicted.

This book tells the story of what we've discovered so far.

The Traditional Theory of Meaning

For thousands of years, scientists and philosophers have been trying to figure out how meaning works. And yet good answers have been awfully hard to come by, much more so than for other aspects of language. The fields of linguistics and psychology have actually made substantial strides in the way that people pronounce and perceive words and the reasons why words in sentences take the particular orders they do. These are aspects of language that are directly measurable—you can tell exactly when a speaker's tongue makes contact with the velum to pronounce a hard *k* sound. But meaning is comparatively harder, because it's something that you do almost entirely in your mind. As a result, it's invisible to direct inspection—we can't measure it, count it, or weigh it. That makes it hard to bring the usual means of science to bear on it. There's no debating the sizeable potential rewards for learning how meaning works, but for most of human history, despite their allure, they've eluded capture. So, although you might expect otherwise, the scientific study of meaning is still in its infancy.

However, even in the absence of solid empirical evidence, theories about how meaning works have developed and thrived. Over the years, most linguists, philosophers, and cognitive psychologists have come to settle on a particular story, which probably isn't so different from your intuitive sense of meaning. When you contemplate meaning in your daily life, it's likely because you're wondering (or perhaps arguing about) what a given word means. It might be a word in your own language: What does *obdurate* mean? (Stubbornly persistent in wrongdoing, in case you were wondering.) How about *necrophagia?* (Eating the dead.) Or *epicaracy*? (Taking pleasure at others' misfortune.) Or it could be a word in another language: What does the formidable German word *Geschwindigkeitsbegrenzung* mean? (Speed limit.) In general, you're probably most aware of meaning when you're thinking about definitions. This is also the starting point for the traditional theory of meaning: words have meanings that are like definitions in your mind.

What would it be like if meaning worked this way? When you think about it, a definitional meaning would need to have two distinct parts. The first is the definition itself. This is a description of what the word means. It's articulated in a particular language, like English, and is supposed to be a usable characterization of the meaning. But there's a second part, too, which is implicit. The definition characterizes something in the world. So *speed limit* (or if you prefer, *Geschwindigkeitsbegrenzung*) actually refers to something that exists in real life, independent of your knowledge about it—whether you know that there's a speed limit, or what it is, you can still get pulled over for driving faster than the number on the sign. So both the mental definition and the actual thing in the world that the word refers to are each critical parts of the meaning of a word.

Many philosophers have taken it as a given that these two parts are all you need to characterize meaning.[3] And they've gone on to argue for centuries about which of the two parts is more important—the mental definition or the real world. But the important question for our purposes—to understand how people understand—is to ask how a definitional theory of meaning like this could explain the things we do with language. Do we really have these definitions in our minds? If so, where do they come from? How could we use them to plan a sequence of words? How could we use them to understand something that someone else has said?

This is where things get a little more complicated. As with any definition, your mental definitions would presumably need to be articulated in some language. But what language? Your first thought might be that it should be your native language, so English words have mental definitions in English, and German words have mental definitions in German. Except, when you follow that idea to its logical conclusion, there's a problem. If English words are defined in your mind in terms of other English words, then how do you understand the definitions themselves? You end up going in circles. Here's an illustration of the problem from a real-life situation that you might be able to relate to.[4] Suppose you don't speak Japanese. But you're at a train station in Tokyo, and there's a sign you want to look up the meaning of. So you pull out your dictionary and look up the characters, but at that point you realize, to your chagrin, that instead of a Japanese-English bilingual dictionary, you accidentally bought a Japanese-Japanese monolingual dictionary. Oops. On the sign, there's a squiggly character with a horizontal line and some dots, so you look that up in your dictionary, but, regrettably, the definition is nothing but a long string of many more characters that you also don't recognize. You could try to look these up, in turn, but you'd just get more of the same. The problem is the same one that you would have if your mental definitions were expressed in your native language. Definitions expressed in a particular language don't mean anything unless you understand the language already. So in understanding the word *polar bear*, say, it wouldn't work to go through a process of activating an English definition of *polar bear* (a large, white, carnivorous bear that inhabits arctic regions) in your mind. This definition wouldn't be any more meaningful than the *polar bear* that you started out with.

One solution to this problem is to suppose that we have some other system in our mind—some way to encode ideas and thoughts and reasoning that doesn't use English or any real language. This mental language would need to have a lot of the stuff that a real language has—it would still have to be able to refer to things in the world, as well as properties, relations, actions, events, and so on—anything that we can think about and understand language about. In other words, we might be thinking using something like a *language of thought* or *Mentalese*.[5] Simply stated, the language of thought hypothesis is that the meanings of words and sentences in any real language are articulated in people's minds in terms of this other, mental language. Mentalese is supposed to be like a real language in that there are words that mean things and can combine with one another, but, unlike a real language, it doesn't sound like anything or look like anything. So, in Mentalese, we have a word that represents speed limits, and another for epicaracy, and another for polar bears, and so on. To understand a real language like English or Chinese, we need to translate the words we hear or read into Mentalese. So the language of thought hypothesis breaks mental definitions out of their self-referential circle by seeing the human capacity for meaning as akin to using a bilingual dictionary instead of a monolingual one. If you showed up at a Japanese train station with a Japanese-English dictionary, you could understand what the Japanese characters meant by looking them up in the dictionary, because the dictionary translates them into words a language you already know. And by analogy, the language of thought hypothesis states that for each word that we know, we have a mental entry that includes a definition articulated in Mentalese. This is one of the most important and influential ideas people have had about meaning and the mind.

But even if Mentalese gets us out of the vicious circle of words defined in terms of other words, it still only gets us part way to meaning. That's because it doesn't deal with the other half of a definitional theory of meaning—the things in the world that the Mentalese words refer to. According to the language of thought hypothesis, the words of Mentalese are related to the world through a symbolic relationship. For instance, when you read the words *polar bear* and translate them into whatever your Mentalese word for polar bear is, let's call it 9us&'~ (as a reminder that it's not supposed to be pronounceable), that word has meaning by dint of the set of things in the world that are actually polar bears. So a sentence like *The polar bear mostly blends in with its icy, snowy surroundings* has meaning because it describes a situation in the world where a thing appropriately designated by your symbol for *polar bear* is in fact doing something designated by your symbol for *blends* into something designated by your symbols for *icy, snowy surroundings*.

Over the centuries, this has come to be the leading idea about how meaning works. Words are meaningful because you have mental definitions for them—articulated in Mentalese—that match up to things in the real world.

Embodied Simulation

But if you look a little closer at the language of thought hypothesis, you'll find that there are actually some holes in it. The biggest one is that Mentalese doesn't actually solve the problems inherent in a definitional theory of meaning—it simply pushes them back a level. The issue is akin to the earlier question of how an English definition of an English word could ever mean anything. Namely: How do we know what the words in Mentalese mean? What language are <u>they</u> defined in? How does activating a sentence in Mentalese actually create meaning? How does it allow us to understand?

One way to think about this issue is using a version of a thought experiment known as the Chinese Room argument.[6] Say you're sitting in an enclosed room with two slots in it. Occasionally, someone will slide a card written in Chinese characters into the room through one of the slots. Now, you don't know any Chinese, but your job is to look these characters up in a book. The book will have some other characters next to the one you looked up, and you're supposed to find a card with those other characters on them and slide it out of the room through the other slot. Because you don't know Chinese, you have no idea what's on the cards, but people outside the room, who do know Chinese, think that the person in the room must certainly be a native Chinese speaker because the responses that come out of the room are perfectly appropriate rejoinders to the messages that they slip into the room. Of course, this is only possible if the book you're looking up the answers in is really well designed. But the question is: Do you understand Chinese? I suspect you'll agree that, no, of course you don't. We can apply the same reasoning to the language of thought hypothesis as an explanation of how meaning works. The Chinese characters in this example are like the words of Mentalese. Simply identifying and arranging symbols in some language, even if those symbols represent something in the real world, isn't enough to make meaning. It's not enough to say you've understood something.

This is one of the big problems with the language of thought hypothesis. And when you start to apply a little pressure, other cracks start to appear. For one, where does Mentalese come from? If it's something that's learned, then it certainly can't be learned through one's native language, because that creates another vicious cycle: How could we learn Mentalese based on English if we only understand English through Mentalese? So if Mentalese can't be learned from language, then that means that—if there is such a thing as Mentalese—it has to exist in our minds before we even start to learn language. In other words, in order to learn the English *polar bear*, we have to already have a Mentalese symbol representing polar bears. And this also means that people who speak different languages must all have the same underlying concepts—a polar bear is a polar bear is a polar bear. There's good reason to question all of these claims.

Even the greatest strength of the language of thought hypothesis—the simplicity of Mentalese symbols—is gained at substantial cost. The idea that the weight of meaning might be carried by Mentalese symbols is quite powerful and appealing, because those symbols would be so simple. Symbols are pointers that just tell you what things in the world they refer to. To understand what the English word *polar bear* means is to have a symbol 9us&'~ that refers to actual polar bears in the world. To understand what the English word *dog* means is to have some other symbol, maybe THX1138. But the only way to allow the symbols to be that simple is to leave out most of the details. The fact is that you probably know a lot about polar bears— their color, how they move, exactly how afraid of them you should be, what type of carbonated soft drink they purportedly prefer around the winter holidays, and so on. That's a lot to know, especially for something like polar bears, which you know comparatively little about. Think about something you know much more about, like dogs. You probably know what they look like (and there's lots of variability here by breed and age) and what they smell like (this also varies, by wetness, recency of rolling in fish, and so on), but also how they evolved from wolves and the fact that they can be recruited to pull sleds and that they take fondly to being scratched above their tail. But a Mentalese word for *polar bear* or one for *dog* would be equivalently simple symbols that refer to the category of polar bears or dogs and skip over all this detailed and variable knowledge. A Mentalese symbol for dogs isn't the collection of memories you have of interacting with dogs or the breed of puppy you're hoping you'll get for your birthday. Instead, it's just a symbol that points to the range of things in the world that are in fact dogs. That's the thing. Symbols in Mentalese are first and foremost symbols. Meaning is simple, clean, logical, and efficient. As a result, there's no place in this theory of meaning for the details.

Clearly, thinking of meaning in terms of Mentalese symbols has some limitations. But until recently, it was the best game in town. Our best guess was imperfect, but we didn't have the right empirical evidence to tell us what was really going on.

That didn't stop at least some people over the years from realizing that the emperor, if not entirely denuded, was revealing some indecent parts. Starting as early as the 1970s, some cognitive psychologists, philosophers, and linguists began to wonder whether meaning wasn't something totally different from a language of thought. They suggested that—instead of abstract symbols—meaning might really be something much more closely intertwined with our real experiences in the world, with the bodies that we have. As a self-conscious movement started to take form, it took on a name, *embodiment*, which started to stand for the idea that meaning might be something that isn't distilled away from our bodily experiences but is instead tightly bound by them. For you, the word *dog* might have a deep and rich meaning that involves the ways you physically interact with dogs—how they look and smell and feel. But the meaning of *polar bear* will be totally different, because you likely don't have those same experiences of direct interaction. If meaning is based on our experiences in our particular bodies in the particular situations we've dragged them through, then meaning could be quite personal. This in turn would make it variable across people and across cultures. As embodiment developed into a truly interdisciplinary enterprise, it found footholds by the end of the twentieth century in linguistics, especially in the work of U.C. Berkeley linguist George Lakoff and others;[7] in philosophy, especially in work by University of Oregon philosopher Mark Johnson, among others;[8] and in cognitive psychology, where U.C. Berkeley psychologist Eleanor Rosch's early work led the way.[9]

The embodiment idea was appealing. But at the same time, it was missing something. Specifically, a mechanism. Mentalese, for all its limitations, is a specific claim about the machinery people might use for meaning. Embodiment was more of an idea, a principle. It might have been right in a general sense, but it was hard to tell because it didn't necessarily translate into specific claims about exactly how meaning works in real people in real time. So it idled, and it didn't supplant the language of thought hypothesis as the leading idea in the cognitive science of meaning.

And then someone had an idea.

It's not clear who had it first, but in the mid-1990s at least three groups converged upon the same thought. One was a cognitive psychologist, Larry Barsalou, and his students at Emory University, in Georgia.[10] The second was a group of neuroscientists in Parma, Italy.[11] And the third was a group of cognitive scientists at the International Computer Science Institute in Berkeley, where I happened to be working as a graduate student.[12] There was clearly something in the water, a zeitgeist. The idea was the *embodied simulation hypothesis*, a proposal that would make the idea of embodiment concrete enough to compete with Mentalese. Put simply:

> Maybe we understand language by simulating in our minds what it would be like to experience the things that the language describes.

Let's unpack this idea a little bit—what it means to simulate something in your mind. We actually simulate all the time. You do it when you imagine your parents' faces, or fixate in your mind's eye on that misplayed poker hand. You're simulating when you imagine sounds in your head without any sound waves hitting your ears, whether it's the bass line of the White Stripes' *Seven Nation Army* or the sound of screeching tires. And you can probably conjure up simulations of what strawberries taste like when covered with whipped cream or what fresh lavender smells like. You can also simulate actions. Think about the direction you turn the doorknob of your front door. You probably visually simulate what your hand would look like, but if you're like most people, you do more than this. You are able to virtually feel what it's like to move your hand in the appropriate way—to grasp the handle (with enough force to cause the friction required for it to move with your hand) and rotate your hand (clockwise, perhaps?) at the wrist. Or if you're a skier, you can imagine not only what it looks like to go down a run, but also what it feels like to shift your weight back and forth as you link turns.

Now, in all these examples, you're consciously and intentionally conjuring up simulations. That's called *mental imagery*. The idea of simulation is something that goes much deeper. Simulation is an iceberg. By consciously reflecting, as you just have been doing, you can see the tip—the intentional, conscious imagery. But many of the same brain processes are engaged, invisibly and unbeknownst to you, beneath the surface during much of your waking and sleeping life. Simulation is the creation of mental experiences of perception and action in the absence of their external manifestation. That is, it's having the experience of seeing without the sights actually being there or having the experience of performing an action without actually moving. When we're consciously aware of them, these simulation experiences feel qualitatively like actual perception; colors appear as they appear when directly perceived, and actions feel like they feel when we perform them. The theory proposes that embodied simulation makes use of the same parts of the brain that are dedicated to directly interacting with the world. When we simulate seeing, we use the parts of the brain that allow us to see the world; when we simulate performing actions, the parts of the brain that direct physical action light up. The idea is that simulation creates echoes in our brains of previous experiences, attenuated resonances of brain patterns that were active during previous perceptual and motor experiences. We use our brains to simulate percepts and actions without actually perceiving or acting.

Outside of the study of language, people use simulation when they perform lots of different tasks, from remembering facts to listing properties of objects to choreographing a dance. These behaviors make use of embodied simulation for good reason. It's easier to remember where we left our keys when we imagine the last place we saw them. It's easier to determine what side of the car the gas tank is on by imagining filling it up. It's easier to create a new series of movements by first imagining performing them ourselves. Using embodied simulation for rehearsal even helps people improve at repetitive tasks, like shooting free throws and bowling strikes. People are simulating constantly.

In this context, the embodied simulation hypothesis doesn't seem like too much of a leap. It hypothesizes that language is like these other cognitive functions in that it, too, depends on embodied simulation. While we listen to or read sentences, we simulate seeing the scenes and performing the actions that are described. We do so using our motor and perceptual systems, and possibly other brain systems, like those dedicated to emotion. For example, consider what you might have simulated when you read the following sentence a little while ago:

> When hunting on land, the polar bear will often stalk its prey almost like a cat would, scooting along its belly to get right up close, and then pounce, claws first, jaws agape.

To understand what this means, according to the embodied simulation hypothesis, you actually activate the vision system in your brain to create a virtual visual experience of what a hunting polar bear would look like. You could use your auditory system to virtually hear what it would be like for a polar bear to slide along ice and snow. And you might even use your brain's motor system, which controls action, to simulate what it would feel like to scoot, pounce, extend your arms, and drop your jaw. The idea is that you make meaning by creating experiences for yourself that—if you're successful—reflect the experiences that the speaker, or in this case the writer, intended to describe. Meaning, according to the embodied simulation hypothesis, isn't just abstract mental symbols; it's a creative process, in which people construct virtual experiences—embodied simulations—in their mind's eye.

If this is right, then meaning is something totally different from the definitional model we started with. If meaning is based on experience with the world—the specific actions and percepts an individual has had—then it may vary from individual to individual and from culture to culture. And meaning will also be deeply personal— what *polar bear* or *dog* means to me might be totally different from what it means to you. Moreover, if we use our brain systems for perception and action to understand, then the processes of meaning are dynamic and constructive. It's not about activating the right symbol; it's about dynamically constructing the right mental experience of the scene.

Furthermore, if we indeed make meaning through simulating sights, sounds, and actions, that would mean that our capacity for meaning is built upon other systems, ones evolved more directly for perception and action. And that in turn would mean that our species-specific ability for language is built up from systems that we actually share in large part with other species.

Of course, we use these perception and action systems in new ways. We know this because other animals don't share our facility with simulation. Returning to the polar bear, I actually have some bad news to share. Since the early reports of their nose-covering behavior, polar bears have been observed a lot—in zoos, in the wild. And despite all the hype, it turns out that there's basically no modern evidence of nose-covering behavior.[13] Sorry to disappoint you. But there is actually a much deeper lesson here. A polar bear, unlike a human, probably can't simulate what it looks like to its potential next meal. The capacity for open-ended simulation is something much more human than ursine, not just in language, but pervasively throughout what we do with our minds. You can simulate what you would look like if you covered your nose with your hand, just as easily as you can simulate what you'd look like if you had two heads or if you had a pogo stick in place of your right leg. If simulation is what makes our capacity for language special, then figuring out how we use it will tell us a lot about what makes us unique as humans, about what kind of animal we are, and how we came to be this way.

Flying Pigs

One of the important innovations of the embodied simulation hypothesis—and one way in which it differs from the language of thought hypothesis—is that it claims that meaning is something that you construct in your mind, based on your own experiences. If meaning is really generated in your mind, then you should be able to make sense of language about not only things that exist in the real world, like polar bears, but also things that don't actually exist, like, say, flying pigs. So how we understand language about nonexistent things can actually tell us a lot about how meaning works.

Let's consider the case of the words *flying pigs*. I'd wager that *flying pigs* actually means a lot to you, even without thinking too hard about it. Over the years, I've asked a lot of people what *flying pigs* means to them, informally. (One of the luxuries of being a university professor is that people tend to be totally unsurprised when you ask questions like *How many wings does a flying pig have?*) According to my totally unscientific survey, conducted primarily with the population of individuals with time on their hands and a beverage in their glass, when most people hear or read the words *flying pigs*, they think of an animal that looks for all intents and purposes like a pig but has wings. The writer John Steinbeck imagined such a winged pig and named it *Pigasus*. He even used it as his personal stamp. What do you know about your own personal Pigasus? It probably has two wings (not three or seven or twelve) that are shaped very much like bird wings. Without having to reflect on it, you also know where they appear on Pigasus' body—they're attached symmetrically to the shoulder blades. And although it has wings like a bird, most people think that Pigasus also displays a number of pig features; it has a snout, not a beak, and it has hooves, rather than talons.

There are a couple things to draw from this example. First, *flying pigs* seems to mean something to everyone. And that's important because there's no such thing as an actual flying pig in the world. In fact, part of the meaning of *flying pigs* is precisely that flying pigs don't exist. What all of this means, not to be too cute about it, is that the Mentalese theory that meaning is about the relation of definitions to real things in the world will only work when pigs fly.

Second, if you're like most people, what you did when you understood *flying pigs* probably felt a lot like mental imagery. You might ask yourself, did you experience visual images of a flying pig in your mind? Were they vivid? Were they replete with detail? Of course, consciously experiencing visual imagery is just one way to use simulation—you can also simulate without having conscious access to images. But where there's imagined smoke, there may be simulated fire. If you're like most people, when you simulate a flying pig, you probably see the snout and the wings in your mind's eye. You may see details like color or texture; you might even see the pig in motion through the air. The words *flying pigs* are not unique in evoking consciously accessible visual detail. The same is true for lots of language, whether the things it describes are impossible like *flying pigs* or totally mundane like *buying figs* or somewhere in between, like *the polar bear's nose*.

Third, and I don't expect that this occurred to you because it only became clear to me through my extensive research—*flying pigs* doesn't actually evoke something of the genus Pigasus for everyone. For some people, flying pigs don't use wings to propel themselves, but instead conscript superpowers. If your flying pig is of this variety—let's call it *Superswine*—then it probably wears a cape. Maybe a brightly colored spandex unitard, too, with some symbol on the chest, like a stylized curly pig tail or, better yet, a slice of fried bacon. And what's more, when it flies, Superswine's posture and motion are different from those of winged flying pigs. Whereas winged flying pigs hold their legs beneath their body, tucked up to their bellies or hanging below them, Superswine tend to stretch their front legs out in front of themselves, à la Superman (see Figure 1).

I'll be the first to admit that the respective features of Pigasus and Superswine are not of great scientific value or vital public interest in and of themselves. But they do tell us something about how people understand the meanings of words. People simulate in response to language, but their simulations appear to vary substantially. You might be the type of person to automatically envision Superswine, or you might have a strong preference for the more common Pigasus. We observe individual variation like this not only for *flying pigs*, but equally for any bits of language. Your first image of *a barking dog* might be a big, ferocious Doberman, or it might be a tiny, yappy Chihuahua. When you read *torture devices*, you might think of the Iron Maiden or you might think of a new Stairmaster at your gym. Variation in the things people think words refer to is important because it means that people use their idiosyncratic mental resources to construct meaning. We all have different experiences, expectations, and interests, so we paint the meanings we create for the language we hear in our own idiosyncratic color.

And finally, *flying pigs* teaches us that when you engage your visual system to understand language, you do so creatively and constructively. You can take previously experienced percepts (such as what pigs look like) and actions (such as flying) and form new combinations out of them. What *flying pigs* means depends on merging together independent experiences, because you have probably never experienced anything in the real world that corresponds to *flying pigs* (unless

FIGURE 1 Artistic depictions of Pigasus (left) and Superswine (right).

you spent a lot of time at Pink Floyd concerts in the 1970s). That makes *flying pigs* an extreme case, but even when language refers to a corresponding real-world entity—even in mundane cases—you still have to build up a simulation creatively. Consider the totally boring expression *yellow trucker hat*. Now, surely there exist yellow trucker hats in the world. You have probably seen one, whether or not you were so moved by the experience as to remember it. But unless you have a specific stored representation of a particular yellow trucker hat, the mental images that you evoke to interpret this string of ordinary words have to be fabricated on the spot. And to do this, you combine your mental representation of *trucker hat* with the relevant visual effects of the word *yellow*. When words are combined—whether or not the things they refer to exist in the real world—language users make mental marriages of their corresponding mental representations.

The New Science of Meaning

The next step is to put the idea of embodied simulation under a microscope and really put it to the test. But how? The currency of science is observable, replicable observations that confirm or disconfirm predictions, but, as I noted earlier, meaning doesn't lend itself willingly to this kind of approach because it's quite hard to observe. So, what to do? Facing this quandary as you are, you're in pretty much the same place where the field of cognitive science was in about the year 2000. There was this exciting, potentially groundbreaking idea about simulation and meaning, and yet we had no idea how to test it.

And that's when the ground shifted. Right about at the same time, a handful of trailblazing scientists started to develop experimental tools to investigate the embodied simulation hypothesis empirically. They flashed pictures in front of people's faces, they made them grab onto exotically shaped handles, they slid them into fMRI scanners, and they used high-speed cameras to track their eyes. Some of these approaches failed completely. But the ones that worked rocketed meaning onto the front page of cognitive science. And they provided us with instruments that now allow us to scrutinize humans in the act of making meaning.

The next ten chapters are a tour through this fascinating new science that is unraveling how meaning works. To make headway toward an answer, we'll look first at how people use simulation when they're not using language, for instance, when they're merely imagining hypothetical situations or recalling past experiences. By spending some time with bowlers visualizing how to bowl strikes and with competitive memory champions remembering randomized decks of cards, we'll discover how people think by simulating the sights and sounds and actions that they're thinking about. We'll then carry this insight over to language and look at the evidence that people do the same thing for sights, sounds, and actions that they hear or read about. Subsequent chapters explore the details: how people understand language about things that they can't see or hear, like *ideas* and *time*, how the grammar of a sentence affects the meaning people extract from it, how meaning differs from culture to culture, and how people with different experiences understand the same words and sentences differently. The product is an account of how people comprehending language avail themselves of the various cognitive systems at their disposal to actively create an understanding of the words they hear. In other words, this is the story of how you ever manage to understand anything. It's the story of how you breathe life into your own personal Pigasus and the story of how you figured out why a polar bear would ever cover its nose.

CHAPTER 2

Keep Your Mind
on the Ball

In 1991, Barry Bonds played outfield for the Pittsburgh Pirates. He was a quick, dynamic player but not a physically imposing one. At 185 pounds and 6'1" tall, he had a similar build to that of his father Bobby, also a professional ball player. Then, in 1993, Barry moved to San Francisco to play for the Giants and something happened. He got big. Over the next ten years, he increased his weight by nearly 25 percent. And this was no typical middle-age ripening of the belly. He added muscle. The slim outfielder became a beefy 228-pound slugger. And it helped his game. From 1991 to 2001, he nearly tripled the number of home runs he hit, and in 2001 he had arguably the best offensive year in the history of the game, destroying the previous home run record with 73.

Maybe the California sunshine was good for him. Or it could have been something in the food—perhaps he discovered that sprouts and avocado have previously unrecognized muscle-promoting properties. More likely though, at least according to common wisdom, he discovered anabolic steroids.

Taking steroids carries a host of undesirable potential side effects. They can cause acne, enlarged breasts (in both women and men), and shrunken testicles (only in men). And they're also illegal. And yet athletes continue to endanger their bodies and careers by taking them. Because steroids work. They allow the athlete to train harder and recover more quickly.[1] And the more you're able to train, the better you'll probably do. To hit more home runs, practice hitting a ton of pitches. If you want to improve your tennis serve, serve thousands of balls. To be a better bowler, get out and roll. The more practice, the better.

So Bonds probably hit the juice. And it worked. But right about the same time, there were a bunch of trainers using a different approach, one that they thought would help improve batters' hitting and golfers' putting and not cause anything to unduly enlarge or shrink along the way.

It started in the 1980s. Certain trainers started wondering whether they could help athletes improve their respective skills using a strategy that didn't involve physical training at all. Their motivation was simple. Actual practice has its limitations. For one, it's costly, in that you need a field or court, equipment, trainers, and so on. What's more, you can injure key body parts—muscles, joints—through overuse. (Hence the appeal of steroids.) So what would happen, they mused, if you had athletes practice less and spend their time visualizing their performance instead of actually doing it? Would imagining shooting free throws improve your actual free throw percentage? Would visualizing perfect bowling form produce more strikes?

So they took some tennis players and had them lay off the serves. They had bowlers lay down their balls. They directed basketball players to lie around in the locker room. And they had them visualize.

At first, they probably met with a whole lot of resistance. After all, how could you actually improve at serves and free throws by doing anything other than practicing more? If you're trying to hit a small green ball over a three-foot-tall net into the corner of the service box sixty feet away, then any reasonable person would practice doing exactly that.

But the thing is, all the sitting around worked. The athletes who spent their time visualizing hit more of their serves in, bowled more strikes, and knocked down more free throws.[2] At least, that is, as long as they visualized performing their skill successfully. If they imagined bowling strikes, they were more likely to bowl strikes than if they didn't visualize anything at all. But if they visualized rolling gutter balls, their bowling got worse than if they did nothing.[3]

Visualizing is now a standard part of sports psychology. It doesn't replace anabolic steroids—it's not clear for instance that Barry Bonds would have reached his peak home run–hitting potential without controlled substances. Nevertheless, visualization works. The question is why—why is it that imagining using your body in a particular way makes it easier for you to move your body in that same way later? The answer that we'll see in this chapter is remarkably simple. When we visualize actions—consciously and intentionally activating mental images—we use the very parts of our brain that control our body's movements. When we imagine the footwork we employ to serve a tennis ball, the part of our brain that controls foot motion starts firing. When we think about how we hold a basketball in our hands, the part of our brain controlling hand motion lights up. As a result, whether you call it mental imagery, visualization, or mental rehearsal, imagining doing things is extremely effective at solidifying motor skills. And that's because, to a large extent, when we're visualizing, our brain is doing the same thing it would in actual practice. Admittedly, we're missing some things that might be quite useful, like the feedback and conditioning we gain from actual action. But at the same time, we also avoid the costs and strains on our body. The important thing is that visualizing an action—in terms of what our brain is doing—is a lot like performing the action.

And the bigger picture that we're going to see in this chapter is that what's true of visualization is also true for a variety of other mental activities. Actively imagining or visualizing an action reuses parts of the brain that actually control those imagined actions. And this is a pervasive property of the mind. We also use our brain's action and perception systems for memory; when we recall events, we reconstruct what they felt like, looked like, or sounded like, and this again uses parts of our brain whose primary duty is to allow us to perceive or participate in events of those types in the first place. And likewise when we think about properties of objects, to decide whether an object has certain properties, for instance, what color a polar bear's nose is, we use our vision system to construct a mental representation of what a polar bear's face looks like. All of this is to say that lots of things we can do with our brain—things other than moving our bodies or perceiving the world—make use of the parts of the brain primarily responsible for moving our body or perceiving the world.

As it turns out, the trainers exploring visualization hit on something that cognitive psychologists have been studying for a long time. At the outset, the psychologists were mostly looking at visual imagery, images in the mind's eye. So that's where we'll start, in a lab at Cornell, at the turn of the twentieth century.

The Perky Effect

In 1910, an innovative, young American cognitive psychologist named C. W. Perky was experimenting with a relatively new technology, passing light through film to project images on a blank wall. Of course, this technology would come to revolutionize entertainment in the form of moving pictures. And it turns out that it did the same for the study of the mind. Perky wanted to explore what was going on inside people's heads while they were performing mental imagery—while they were actively, consciously conjuring up images of things that weren't in fact in front of them.[4] What she did was to ask participants to imagine seeing an object (like a banana or a leaf) while they were looking at this blank wall. Meanwhile, unbeknownst to them, she projected an actual image of the object they were supposed to be imagining on the wall. At first, the projected image was below the threshold for the participants to consciously perceive it, but she increased the illumination slowly so that it became more and more visible. Perky found that many participants continued to believe that they were still just imagining the banana or leaf and failed to recognize that there was actually a real, projected image on the wall. And yet she knew that the projection was visible, because when she asked other participants who were not performing imagery to look at the wall, they were able to see the green or yellow image easily.

Perky's experiment showed that performing mental imagery can interfere with actually perceiving the world. The so-called *Perky effect* comes up a lot in daily life. It happens, for example, when you daydream. During daydreaming, you're completely awake and your eyes are eyes open, and yet you're imagining being somewhere else, doing something else, seeing things that aren't there. Maybe you're disappointedly picturing the contents of your fridge and wondering how you're going to conjure up a respectable dinner out of half a cup of bacon bits and a bottle of ketchup. Or you're envisioning the look on your boss' face when you tell her you've won the lottery and are moving to Hawaii. But while you're doing all this, despite having your eyes open, you aren't processing much of the visual world around you. If you're in a classroom, you don't see what's being written on the board. If you're in a car, you might not see the cars around you or the street signs (you just missed your exit, by the way). Imagery interferes with vision.

Naturally, research on the Perky effect didn't stop with its discovery in 1910. More recent studies have started to paint a somewhat more textured picture of how imagery interacts with vision. For one, what you're imagining affects whether it interferes with what you're seeing.[5] If you have someone look at the middle of a computer screen and imagine a "T," and then you display either a "T" or an "H" in the middle of the same screen, what you find is that they are less likely to correctly detect the displayed letter when it's actually an "H" than when it's a "T" (and vice versa if you have them imagine an "H"). Here's how that might play out in your real life, if it's anything like mine. Suppose you're looking for your brand new phone, which is red, you're a little distracted, and as a result you're actually holding in mind a mental image of your old phone, which was black. Because you've got in mind a mental image of a black phone, it will be relatively hard for you to actually see your new, red phone even if it you look right at it. But as soon as you remember that you have a new, red phone your mental image no longer interferes with actually detecting the phone when you in fact come across it.

The story gets more involved, because it's not only *what* you're imagining that affects whether imagery will interfere with perception but also *where* you imagine it being.[6] Suppose you're seated in front of a computer screen and told to imagine an object. Say, the capital letter "I." Then an asterisk appears somewhere on the screen, and you have to push a button as soon as you see it. The Perky effect would predict that it should be harder for you to detect the asterisk on the screen when you're imagining the "I" than when you're not—because the "I" and the asterisk look different. But it turns out that this is only the case when you're imagining the "I" in the same location as the asterisk is presented. That is, if you're told to imagine the "I" in the upper part of the screen, then you're less likely to accurately detect an asterisk in the upper part of the screen. But when the locations are different—when you imagine an "I" in the upper part of the screen and then are shown an asterisk in the lower part of the screen—you detect the asterisk just as well as you do when not performing imagery at all.

This line of Perky-inspired research shows that although visual imagery sometimes interferes with vision—vision of different things in the same location—it also sometimes can actually enhance vision.[7] When the image depicts the same thing you're imagining and it's in the same place and of a compatible shape and size, you are actually better at perceiving the thing in front of you than when you're not performing imagery at all—if you're (correctly) envisioning your new, red phone, and you cast your eyes over it, you'll see it more quickly than if you had no image of it at all.

What to make of the fact that imagery and perception interact? There could be two things going on. The first is that visual imagery and visual perception might be performed by the same parts of the brain, which would cause interference because it's known that we can't use the same brain tissue to do two different things at the same time. This is an intriguing idea—that we use parts of our brain that perform vision to also "see" in our mind's eye. Alternatively, though, it could be that our capacity for imagery and vision are just performed by two very tightly linked but distinct systems, and that when these systems are working on the same problem, they thrive, but when they aren't in sync, they don't work so well. And although it doesn't matter to us when we can't find our cell phones why it is that our brains don't let us see them when they're right in front of us, it matters to anyone who wants to know how mental imagery works.

Fortunately, there's a second finding that Perky reported in her original study that helps us distinguish between these two possibilities. When Perky asked her experiment participants to describe their mental images of bananas or leaves, they reported imagining objects that conformed to the shape and orientation of the images she had been projecting on the screen, which is quite surprising, because they had said that they hadn't seen the images themselves! So if she displayed a vertically elongated yellow blur while her subjects were imagining a banana, they reported having imagined a banana that was standing upright. A horizontal yellow blur would produce reports of horizontal bananas. In other words, even though they didn't report actually seeing a projected banana, Perky's participants nevertheless integrated what they were actually seeing projected onto the screen with the mental imagery they were constructing.

More recent research has shown in an even more lucid way how people integrate what they're imagining with what they're seeing. In one study, people were asked to imagine the New York skyline while looking at a screen.[8] Meanwhile, the experimenters projected a faint red circle on the screen. The projection was faint enough (and the Perky effect was strong enough) that the participants didn't report seeing the red circle. However, some of them reported that they had constructed a mental image of New York at sunset.

What these last findings show is that the things you see can be confused with—and integrated with—the visual images you fabricate. The simplest way to understand these facts is that visual imagery and visual perception are subserved by brain systems that are more than merely linked together. For vision to be confused as imagery—for people to actually think that what they're actually seeing is merely the content of their imagination—there must be at least some overlap in the brain systems that people use to perform the two distinct behaviors. Later in this chapter, we'll look at brain imaging studies that will allow us to see exactly how much overlap there is between seeing and imagining and where in the brain it's located.[9]

The Mind Spins

If the same brain systems are used to imagine and perceive objects at rest, like bananas and leaves, then it's quite possible we could also use the same systems to both imagine and perceive objects when they're in motion. One of the classic studies in cognitive psychology showed this in a particularly compelling way.[10] Look at each of the pairs of objects in Figure 2. Each pair of objects (for example, the two shapes in A) is either the same object, rotated differently, or two different objects that are mirror images of each other. Your job is to decide as quickly as possible whether they're the same object, or mirror image objects.

So are the two objects in A the same object, just rotated, or mirror image objects? How about B or C? (If you want the answers, check this endnote.[11])

It's worthwhile to reflect for just a moment on how you figured it out. What psychologists find in this experiment and others like it, whether using letters[12] or complex shapes,[13] is that the time it takes people to answer that two images are rotated versions of the same object increases linearly with the degree of rotational difference between them. That is, if it takes you two seconds to make a decision for objects that are off by forty degrees, and three seconds for objects that

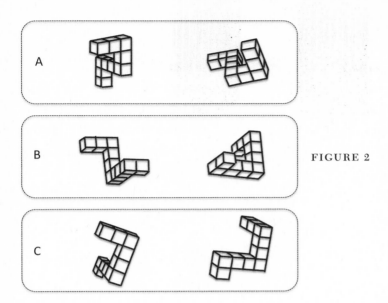

FIGURE 2

are off by sixty degrees, then it will take you four seconds for objects off by eighty degrees, and so on. The best explanation for this finding—and the intuitive answer most participants give when asked how they compared the objects—is that you're imagining rotating one object to see if it lines up with the other one. Mental rotation, as it turns out, works at a constant rate—measured in mental degrees per second—so it makes sense that it takes you progressively longer to respond the farther you have to rotate your mental image of the object.

There's another analogy between perceived motion and imagined motion. Suppose you're up in a helicopter, looking down at an island below you, maybe one that looks like Figure 3. You see a hut, a well, some trees, a lake, grass, a beach, and maybe other landmarks. If you're looking at the hut in the south, say, and then want to check out the well, it doesn't take you very long at all to move your eyes there, compared, say, with shifting your gaze from the hut to the tall grass well to the north. The farther you have to move your eyes, the longer it takes, and as with mental rotation, the relationship between distance and time is linear—longer motions take proportionally longer than shorter ones. So now suppose you're not looking at the island,

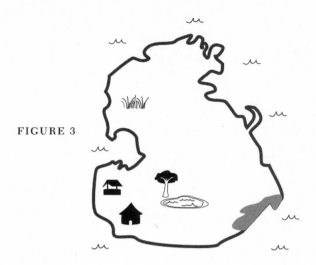

FIGURE 3

but instead merely inspecting a mental image of the island as you re-call it. If you're focusing in your mind's eye on one landmark, say the hut again, and then shift your gaze to the well, it again takes you less time to get there than it does to get, say, to the tall grass. In fact, the relationship between the "distance" your eyes travel in your mental image and the time it takes to get there is again linear.[14] Scanning a mental image is functionally like scanning the real world.

What this all means is that you perceive motion in mental images like you do real motion in the world—the things that take longer in the world also take longer in the mind. Because it's like real motion, mental motion is useful. For example, suppose you want to plan a trip across the island, hitting several key points. If you first imagine a couple different paths, you can get a pretty good estimate of how long each would take, which will save you time when you actually hit the ground. Mental imagery saves you time by letting your mind do the walking.

Stimulating Sounds

Neuroscientists have known for more than a century that specific brain regions are responsible for particular cognitive functions. The phrenologists of the late 1800s believed that not only were different aspects of the human psyche localized at different points within the brain, but that, moreover, the shape of a person's skull served as a measure of how developed those different brain regions were—and thus were a measure of the person's psychological disposition.[15] Although the last part has turned out not to be true—you can't judge a brain by its cover—there is truth to the first part. Specific parts of the brain are most adept at performing particular functions.

Never was this demonstrated more strikingly than in the work of neurosurgeon Wilder Penfield. Penfield worked with people suffering from severe epilepsy—people whose debilitating seizures endangered their well-being and that of those around them. The treatment he pioneered in the 1930s was as revealing as it was shocking—and I mean both of those things literally. He would first surgically expose the patient's brain. The image in Figure 4 shows what this procedure looked like, with a patient identified as D. F.[16] He then searched the brain for the offending region—the damaged area responsible for the epileptic attacks—by applying a targeted but gentle electrical current to various parts. The catch is that in order for him to be able to determine what function each part of the brain performed, he needed patients to be awake, so, during this procedure, they were under only local anesthesia. Penfield would apply the electrical stimulation to a given area, like the numbered ones in the image, and would observe the patient's reaction.

What he found was that—consistently across patients—the same brain regions produced the same sorts of effects. For instance, stimulating the areas to the right of the vertical dotted line in the figure, like those labeled 11, 2, 12, and 13, caused the patient's body parts to twitch or move. He had found the *motor cortex*—the area responsible for sending electrical signals to the body's muscles. Stimulating the areas directly to the left this dotted line, like 10, 8, and 16, caused the patient to experience tingling or numbness in different parts of the body. This is the *somato-sensory cortex*, where the brain collects information about the sense of touch from the skin and muscles.

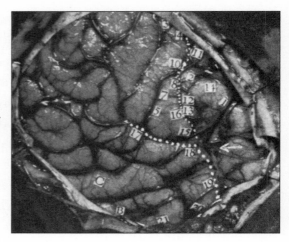

FIGURE 4
Exposed brain of
patient D. F.

But perhaps most surprising was what happened when he applied the electrical current to a spot right between area 21 and area 18. When he asked patient D. F. what she was experiencing, she reported that she heard music. And not just any music—she heard an orchestra playing a specific piece of music, as clearly, she said, as if she were hearing it on the radio. Every time Penfield removed the electrical stimulation and then reapplied it, she heard the same piece of music, starting from the same place, at the same tempo. Other patients had similar auditory experiences with stimulation of this same part of the temporal lobe, though not the same song. Penfield had found one of the brain regions dedicated to audition. When we hear actual sounds, the signal passes through this area, and when electrical current is applied to it directly, it recreates the experience of hearing.

Fortunately for people interested in understanding how auditory imagery works, it can also be induced by far less invasive procedures. For instance, one recent study used an ingenious approach.[17] Participants listened to pieces of music, like the Rolling Stones' "Satisfaction," or the theme song from *The Pink Panther*. Meanwhile, the blood flow in their brain was scanned using an fMRI machine. (The fMRI machine is a tool that allows us to track hemodynamic activity in the brain, which indirectly tells us where neurons are firing while people perform whatever tasks are asked of them.) This enabled the researchers to identify what parts of the brain were at work during listening. But, because they were interested not in how people hear but how they imagine hearing, there was a twist. The experimenters replaced short segments of the music—two to five seconds at a time—with silence. If you've ever driven through a tunnel while listening to the radio, you know that when you're listening to a song you know, as soon as the music cuts out, you spontaneously "hear" the music in your mind's ear over the crackling of your radio. The brain activity measurements that the experimenters took from the periods of silence showed, as you might expect from Penfield's discovery, activation in the brain areas responsible for audition—including the area that, when stimulated, created an experience of musical imagery in patient D. F. The exact parts of the auditory system that were active during the periods of silence depended upon how familiar the music was to the participant and whether it had lyrics—just as you use different but closely related brain regions to hear different types of sound, so you use different brain regions to imagine sound. In recent years, other studies have replicated the same basic finding with a more direct technique. When people are explicitly instructed to imagine specific sounds, auditory brain regions once again light up.[18] The upshot is that we imagine sounds using the same brain regions that allow us to hear real sounds.

Imagery Is Handy

So we've seen that brain systems used for vision and hearing are reused during, respectively, imagery of sights and sounds. But let's return now to actions. Consider some actions that you perform regularly. How much do you know about them? For instance, when you write with a pencil, do you use your ring finger to hold it in place? Before you move on, decide on an answer. Here's another question: Which way do you turn the key to open your front door? Again, try to settle on an answer before moving on.

If you're like most people, you answer these questions in one of two ways. One, you might fool around with your hands, pretending to hold a pencil and turn a key until you can actually see and feel the answer. That is, you might <u>enact</u> the actions you're thinking about. But you might have better things to do with your hands, like maybe hold this book, or maybe you're just lazy and can't be bothered to move, and, so, instead of engaging the muscles of your hand and arm, you might just imagine holding a pencil or turning a key. In other words, maybe you construct *motor images*. It's hard to describe what a motor image is like. Most people have a good intuitive feel for visual images—all I have to do is say *flying pigs* and you're off. But motor images are pretty hard to grasp, so to speak, because they don't look like anything. That's why I asked you about the pencil and the key. Some people report that motor imagery feels like a tingling in the muscles they're imagining moving, while others report some psychological discomfort at both trying to move and holding themselves back. Still others report more somato-sensory than motor imagery. Somato-sensory imagery is an internal re-creation of the feeling of your body—like pressure on your skin, or motion or tension of your muscles. But all these different conscious experiences derive from motor imagery.

In the same way that auditory imagery is performed by brain systems primarily dedicated to hearing, motor imagery uses the parts of the brain that move your body around. Some of the most convincing evidence comes from brain imaging studies. Participants first learn to perform a specific action, such as pressing their four fingers on one hand against their thumb in some order (like index-middle-ring-pinky-index-middle-ring-pinky, etc.). And then they learn to imagine performing the same action without actually moving their body at all. They're then slid into a brain imaging device, for instance an fMRI machine, and perform the action itself and then motor imagery of the action. The experimenters then determine the brain areas that are significantly more active during these activities than some control task (such as imagining looking at a familiar landscape). What they find is that both performing actions and imagining actions engage the key area responsible for sending signals to the muscles in order to move them—the *primary motor cortex*, seen in Figure 5. Perhaps not unexpectedly,

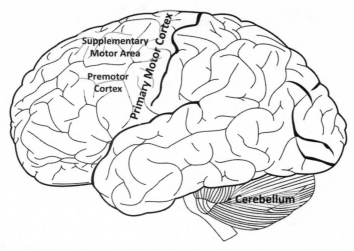

FIGURE 5 Regions involved in motor control, shown on the left hemisphere of the brain.

the activation in this region is often stronger for actual action than imagined action, which in a way is a pale shadow of the real thing.[19]

The thing is, the primary motor cortex isn't homogenous. It's organized topographically, by body part. In essence, there's a map of your body on the surface of your brain. The catch is that the body is represented disproportionately in the motor cortex—those areas that you have finer control over have a larger dedicated portion of the motor cortex. You can find an artistic rendition of this organization in Figure 6. As you can see, the hand, face, mouth, tongue, and larynx take up much more room than the physical size of these body parts size might suggest.

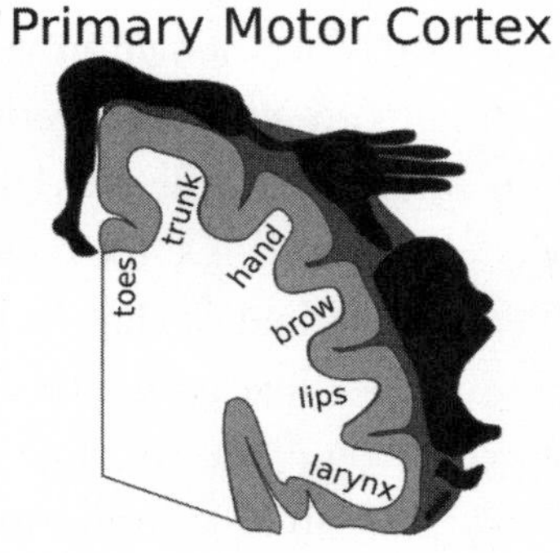

Primary Motor Cortex

FIGURE 6

This organization of the motor strip according to body parts turns out to be quite useful for brain imaging studies. If we have people perform motor imagery in a brain scanner, we can tell whether the specific parts of the motor strip that are active are the ones that are dedicated to performing particular actions using the particular part of the body that the person is imagining. One study had participants first flex their hands, feet, or tongue in the scanner. This allowed the researchers to map out the individual participants' motor strips. Then the participants were asked to just imagine performing these same actions.[20] As in the study described above, the findings showed that primary motor cortex was active during both real action and imagined action, but the intriguing new finding was that this motor cortex activation was body-part specific. That is, when the participants imagined flexing their feet, the top part of their motor cortex lit up (the part that controls foot actions!), but while they were imagining moving their hand or tongue, the middle and lower parts of the motor cortex became active, respectively.

Now, as you saw when you tried to figure out which way you turn the key to your front door, it's not easy to keep your body from moving when you're imagining acting. Could it be that the results from these two studies are due not to motor imagery, but to actual, though unintended, movement? Maybe when they were supposed to be merely imagining actions, participants were inadvertently moving their hands, feet, or tongue. To handle this very reasonable objection, another study used feedback to make sure participants weren't moving at all.[21] The experimenters first taught participants to make a fist and then release it, repeatedly, once per second. Then they trained them to simply imagine performing that action. And they came up with a very clever way to ensure that while the participants were engaging vivid motor images, they weren't moving their hands at all. They used something called electromyography—basically just a readout of the electrical activity of the muscles. They hooked up some electrodes to the participants' hands and showed them electromyographic readouts of what their hand muscles were doing in real time. So the participants could literally see from the fluctuation of the electromygraphic readout whether their fist-clenching muscles were moving at all, even slightly, while they were imagining making a fist. Using this feedback, the participants were able to train themselves to create motor images of fist clenching without actually engaging their hand muscles at all. Once they were good at this, the experimenters used fMRI to locate what parts of the brain the participants were using to actually clench their fists and what parts they were using to imagine clenching their fists. They found three key areas of overlap, and, importantly, both imagining and enacting the actions engaged the main brain regions that control action. What this shows is that motor imagery involves the use of the motor cortex, even when people are definitely not moving.

Now you might be wondering: If you use the motor neurons in your brain to imagine performing actions, then how is it that you're not in a constant state of full-body spasm during the day every time that you imagine physical action? Well, for one thing, the motor activation we observe during motor imagery is never as strong as that during actual action. As a result, it's a weaker signal that gets sent to your muscles, so there's less danger of acting out the actions you merely think about. But in addition, the last study actually made an intriguing finding about one difference between imagery and action. A number of parts of the brain associated with action lit up during both tasks, including the primary motor cortex, which we've been discussing, and others that organize higher-level motor control, like the premotor cortex and supplementary motor area. But one difference between acting and imagining action stands out. That's the cerebellum—the "little brain" in the lower back of the brain—which was active during action execution but not action imagery. This difference is important because the cerebellum plays a role in the coordination of movements (though its exact function is still debated). One way that motor imagery might differ from motor action is that the absence of activity in the cerebellum might shut off actual action, so that the other motor areas can go about imagining performing actions without the body flailing about.

Memory

Because imagery is built upon brain machinery that we use constantly to perceive and act in the world, it has a certain reliability. And, as it turns out, this makes it extremely useful, and not just for straightening out your golf putts. The ancient Greeks discovered an ingenious way to conscript imagery to service. You see, the ancient Greeks—the philosophers, anyway—really liked talking. They told stories, stories that were very, very long. Think Homer. They liked history, debate, philosophy. But the thing is that most of them couldn't read or write. And even those who were literate still didn't necessarily find it to help that much in public speaking. Suppose you wanted to be able to whip out your stirring lecture on why ether is the most ideal of the five elements. As an ancient Greek, you didn't have the luxury of carrying around a Blackberry or iPad that you could load your speech on, and it wasn't particularly convenient to lug around a pile of scrolls for any eventuality. And as a result, the ancient Greeks needed ways to remember lots of stuff without visual aids.

One of their most successful memory innovations was something that we now call the *method of loci*. Here's how it works with something simpler than a Homeric epic poem or treatise on ancient physics. Suppose you need to remember a random sequence of ten words, like the following:

> *water*
> *painting*
> *knife*
> *forest*
> *heart*
> *coffee*
> *boat*
> *nose*
> *radio*
> *key*

You could try to memorize it by rote, repeating the words over and over until it's automatic. And that might work with some effort. But the method of loci works differently and much more deeply. You start by imagining an environment that you know really well, for example, your home. Your home has a lot of different places, like the front door, the kitchen, the bedroom, the closet, and so on, and you can probably imagine what it would look like if you were to stand in each of those places. So now imagine following a predictable path through your home, and, at each salient location, imagine seeing the next item in the sequence of words there. So imagine seeing a glass of water in the entryway. Then you move to the foyer and see a painting (it helps if the objects are particularly vivid, emotionally charged, even lurid). Then you come to the living room, where you see a knife on the table. And so on. If you know the place well enough, if the path you pick is predictable enough that you can reproduce it reliably, and if the associations you make between objects and their locations are vivid enough, you will be able to pretty easily reconstruct the list of words in the very same order you memorized them in, just by using imagery to mentally follow the same path.

Of course, if you're memorizing something more abstract than a list of words, you'll need to come up with images for each important part. But the method of loci presents a very clear example of how mental imagery can be profoundly useful for recalling things.[22] So much so that the world's best modern memorizers use variants of this same method to compete at the World Memory Championships, which is the Olympics of remembering useless information.[23] The reason that the method of loci works so well is that mental imagery is predictable. Visual imagery works much like actual perception because when you recall objects, locations, events, and so on, you are re-experiencing sights you've seen and actions you've performed, using the same brain systems that were responsible for seeing those sights and performing those actions in the first place.

Embodied Simulation

Imagery is useful and enlightening. But at the same time it's quite a specialized cognitive ability. Imagery is intentional—you can make yourself do it if you want. It's conscious, unlike the vast majority of what your brain does. And it's sort of a niche ability, not something you do as frequently as other routine tasks you do with your mind, like thinking about concepts, reasoning, remembering, and so on. So how representative is mental imagery of the way the mind works in general? Sure, people use their visual system when they consciously, intentionally imagine visible things, but what about when they're just thinking about objects, not intending to perform imagery at all? Sure, people can use imagery to remember long lists of things, but what about when they're just trying to remember where they left their keys? Is there an unconscious, unintentional, and pervasive analogue to imagery?

In a word, yes.

Because we've just been talking about memory, let's start there. When you recall things, even when you're not intentionally and consciously performing mental imagery, you reactivate the same brain circuits that you originally used to encode the sights, sounds, smells, and feel of the memories in the first place. We know this from brain imaging studies.

For example, an early study using fMRI had people memorize twenty words, each associated with either a sound or a picture.[24] One person might have had to memorize the word *cow* along with the sound of a cow and *rooster* along with a picture of a rooster, while another participant would memorize the reverse. The next day, the participants were placed in the fMRI scanner and were presented with the words again, but without a picture or sound this time. In the scanner, the participants had to push one of two buttons to indicate, for each word they saw, whether they had previously memorized a sound or a picture for that word. The idea was to measure what parts of the brain became active when they recalled the sounds or the pictures—

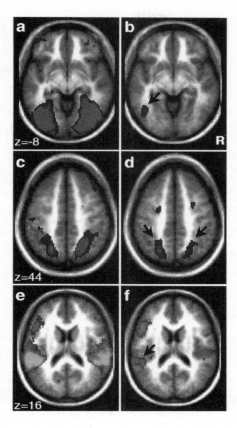

FIGURE 7

whether the brain regions active while people were recalling sounds or pictures were the same as the ones that were responsible for actually hearing or seeing the stimuli in the first place. So to localize these parts of the brain, they then played all the sounds and showed all the pictures to the participants while they were lying in the fMRI machine. The question was whether the people used the same parts of the brain when listening to the sounds and remembering the sounds—whether they used the same parts of the brain when seeing pictures as remembering pictures.

The results are striking, as you can see from the brain images in Figure 7. In each of these images, the back of the brain is at the bottom of the image. On the left, you see the brain areas that are active during perception. The dark splotches in pictures (a) and (c) are the active areas during picture perception—you can see that these are predominantly in the back of the brain, in the occipital lobe, which is responsible for vision. (The top row and second row show the same measurement at different depths in the brain.) The light patches in picture (e) are the areas active during sound perception—these are mostly on the sides of the brain, in the temporal lobe. Now contrast the pictures on the left, which show the regions active during actual perception, with those on the right, taken during the recall task. You can see that the activated regions on the right are parts of the active brain areas for the respective tasks on the left. In other words, recalling pictures uses a portion of the areas that perform actual vision, just as recalling sounds activates part of the region dedicated to hearing sounds. And this is found in a task in which people are just remembering whether they saw a sound or a picture—when they weren't asked to perform mental imagery at all.

Just as recalling sights and sounds activates perception-specific brain areas, so recalling actions activates those parts of the brain that are responsible for engaging those same actions. Here's one Positron Emission Tomography (or PET) study—another form of brain imaging—that shows this quite clearly.[25] While in the PET scanner, participants heard descriptions of actions, for instance *make a fist*, and had to perform them. Then, in a second round, they heard the verbs (in this case *make*) and had to say the noun (*fist*) that had gone with it. The brain imaging data showed a number of areas that were selectively active when people were performing the actions, which you can see from the image on the left of Figure 8. There's a large region of activation toward the top of the brain. The back of this region—toward the left of the image—is the somato-sensory cortex, which detects touch and motion from the body. The front of this blob is the motor cortex, which, as we discussed earlier, is responsible for sending electrical signals to the body's muscles to make them fire. Notice that during recall, in the image on the right, parts of this same blob are also active. You can also see that there are several islands of activation spreading forward and down from this region. These are in regions that are also mostly responsible for coor-

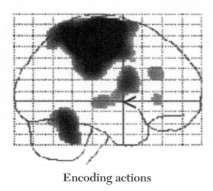
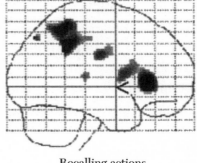

Encoding actions   Recalling actions

FIGURE 8

dinating physical actions, and you can see that the areas active during action and recall are quite similar.

The remarkable thing that these studies show is that recalling facts, like whether a word was paired with a picture or a sound or what noun went with an action verb, uses brain systems dedicated to perception and action. This is important because it suggests that the reuse of the brain systems for other cognitive functions might not be limited to intentional, conscious mental imagery. Instead, it might be an organizing principle of how the mind works.

This is the idea behind the embodied simulation hypothesis. Many of our mental capacities are attributable to simulation. Conjuring up a mental image is a way to consciously and intentionally access an embodied simulation. But behaviors like memory, and others as we'll see below, also use mental simulation. They do so in a more covert way—we may not even be aware that we're mentally simulating. That's because, like most of the rest of what the brain does, embodied simulation isn't necessarily intentional, and it isn't necessarily available for conscious introspection. But it can nevertheless be revealed using some of the same tools of scientific experimentation that revealed its role in imagery. So the embodied simulation hypothesis leads to a pretty clear and testable prediction. In cognitive behaviors other than imagery and recall, people should be using their perceptual and motor systems for simulation— to recreate perception and motor control experiences.

Do Gorillas Have Noses?

We can test this prediction by turning to other run-of-the-mill cognitive tasks and asking whether people appear to be using simulation here as well. For example, one of the core aspects of human cognition is that we know things about objects—we know not only what they look like and how to use them, but, more generally, we know what properties they have. A common task in cognitive psychology experiments that addresses this type of knowledge asks people to make judgments about whether certain objects have particular properties. For instance, does a gorilla have a nose? Does a pony have a mane? Now it's quite possible that, to answer questions like this, you use your vision system to mentally simulate the object and use that embodied simulation to try to find the described property. And you could be doing this totally unconsciously and unintentionally.

Here's a clever way to tell whether this is in fact the strategy that people adopt. If we actually use embodied simulation to detect gorilla noses or pony manes, then the easier the property is to detect visually, the easier it should be for us to determine that the object has that property. What makes properties easier to detect visually? Well, the most obvious thing is their size. A large part of an object is easier to see than a small part—a gorilla's face is easier to see than its nose. So if test subjects verify that a gorilla has a face more easily than that it has a nose, this suggests that the means by which people are arriving at their decisions is vision-like. One study gave people large and small parts of objects to verify and measured both how long it took and how accurate they were.[26] As the embodied simulation hypothesis predicts, the size of object parts was a strong predictor of how quickly and accurately people could verify them. All other things being equal, people are faster and more accurate at confirming large parts, like faces, than small ones, like noses.

Here's another piece of evidence. Suppose you're again asked to verify whether objects have certain properties. For instance, you're given a pair of words like blender–loud, and you have to say whether the second is a property of the first. Now, what might affect how long it would take you to make your determination? One study hypothesized that if thinking about the properties of objects really does engage the specific perceptual systems that those properties pertain to—that is, if determining whether a blender is loud involves performing auditory simulation—then you should verify features about sound more quickly if you're already thinking about sound.[27] Here's how this was tested. The experimenters made a list of object-property pairs. Each of the properties pertained to a single modality: that is, sound, vision, taste, smell, touch, or motor control. And then they manipulated the order in which they presented these object-property pairs. For some participants in the experiment, blender–loud followed an object-property pair in the same modality, like leaves–rustling, which is also implicitly about sound. For other participants, it followed a pair in another modality, like taste: cranberries–tart. And they measured how long it took people to decide that an object had a feature in these two conditions—when it followed a pair using the same modality versus a different modality. What they found was that it took people longer to say that a blender was loud when they had just decided that cranberries are tart (different modality) than when they had just determined that leaves rustle (same modality).

So it seems that routine mental activities, like deciding whether a gorilla has a nose or whether blenders are loud, engage the specific parts of the brain dedicated to the different modes of perception and action. Simulation abounds.

Mental Practice Revisited

So, let's come back to where we started. The remarkable success that athletes experience when they practice by just using their heads should now seem less surprising. The mental machinery that we use to think about bowling or putting is the same machinery that we use to actually perform those same actions and to gain perceptual feedback about how those actions are working. When you imagine bowling, your brain thinks, in a way, that it's actually bowling. This explains not only why you improve performance by mentally practicing but also why only practicing good form yields positive results, while practicing failed technique produces decreases in performance. Imagery, like memory, property verification, and other complex cognitive abilities that we're rightly proud of, are bootstrapped off of evolutionarily older brain systems that allow us to perceive the world and act in it.

This shouldn't be too surprising. Evolution in many ways is a tinkerer—the best biological tinkerer we know of. But it would really be a very poor tinkerer if, given that it had been perfecting a complete vision system for tens of millions of years—and motor and auditory systems for even longer—it then decided to go back to the workbench to construct entirely new, independent machinery for <u>thinking about</u> seeing, hearing, and acting. The use of visual, auditory, and motor imagery for other cognitive functions is the inevitable product of the pressures of efficiency, the limits of mutation, and the demands of ecology. With perception and action systems already in place, how could natural selection help but build a system for other cognitive functions on top of and integrated with these systems?

CHAPTER 3

Meaning and
the Mind's Eye

As animals go, we're heavily biased toward collecting information about the world through our eyes. Dogs rely deeply on smell, and echo-locating bats navigate through sound, but we humans, more like birds of prey and honey bees, prioritize our sense of sight. We prize and value our sight over our other senses (if you ever spent a summer at camp as a child, you probably played out the would-you-rather scenario of which sense you'd prefer to lose, over a belly full of s'mores, and most certainly didn't chose to lose your vision). Vision is the main way we collect information from the world, so it's fitting that vision is also the sense we most closely associate with the internal life of our minds. We even encode it directly in our language—when we talk about meaning, and about understanding, we actually use language about sight. We say: *You see what I mean? The argument was crystal clear? Let's shine some light on this topic.* As we'll see (there it is again!), vision has everything to do with understanding.

We Talk Simple

We humans are endowed with a remarkable capacity for abstract thought. Some humans, now long passed, had the clarity of vision to see how democracy might be a good idea. Some other humans figured out how transfinite numbers work. Still other humans figured out how to put together sequences of musical notes that would make other humans feel something. Yet, despite all our conceptual potential, we spend most of our day intellectually slumming it. Throughout our everyday lives, we mostly think and talk about mundane things we can see and touch. We ask whether you think the milk looks a little yellow after sitting out all night. We systematically talk through which side of the house would be the most dignified burial location for our beloved pet python. And we talk about this awesome car chase in the latest Nick Cage movie where he literally drives a four-wheeler up to the top of the Golden Gate Bridge and then jumps off, and you've got to see it. Why we dedicate our precious seconds on earth to topics like spoiled milk and fictitious car chases is a matter for another book, perhaps one in the self-help aisle. For our present purposes, the point is just that we spend a lot of time communicating about visible things—things that we can presently see, things that we have seen, or things that we expect to see.

And we're quite good at it. With just a few words, I can convey to you a colorful canvas of visual information. Suppose I tell you that *The milk got left out on the counter overnight*. You know that—if you live in a place with a sufficiently warm climate as I do—the milk may take on a characteristic yellow color. If I tell you that *I think we should bury William Snakespeare next to the petunias*, you have a clear idea about the size and shape of the hole that I envision digging. And finally, if I say that *The polar bear mostly blends in with its icy, snowy surroundings*, you know what parts of the bear would stand out against the background. What's amazing is that you know the rough color of the milk and the bear and the approximate size of the snake grave even though the words you read told you nothing of the sort explicitly. This is arguably the most powerful feature of language. Using a limited, discrete code—for instance just the ten or so words in any of the aforementioned sentences—we communicate innumerable subtle details about any arbitrary topic of choice. How do we do this?

One possibility is embodied simulation. We use our visual system not only to detect visible things in the real world but also to mentally simulate nonpresent things. We use visual simulation for certain higher cognitive functions, like recall and categorization. So it seems reasonable to hypothesize that we might also engage our visual system to understand language about visible things. To understand a sentence like *The milk got left on the counter overnight*, you might construct visual representations of a counter—whether a specific one or a generic one—with a container of milk sitting on it, turning a color appropriate to the length of time it's been sitting out. When you hear *I think we should bury William Snakespeare next to the petunias*, you might simulate a snake grave of a particular, appropriate size and shape, possibly in the process of being dug, in dirt, next to a flower bed. In each case, the hypothesis goes, when you are confronted with sentences about visible things, you perform embodied simulations of the events they describe—using your brain's vision system.[1]

Vision 101

When your eyes are open, signals arrive at a specialized brain region dedicated to visual processing called the *primary visual cortex* at the very back of your brain, having traveled all the way from your eyes at the front of your head. The primary visual cortex is the shaded region on the far right side of the brain in Figure 9; it contains neurons that identify basic visual features of what you're seeing—like dots or lines in an image. It then passes this information on to other brain areas for further processing. Activation travels along two separate pathways, one that identifies what objects you see (the *What pathway*) and a second that figures out where they're located (the *Where pathway*). The What pathway is the set of brain systems that compute visual

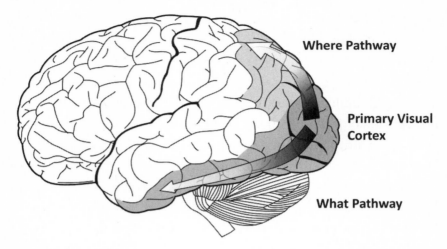

FIGURE 9

properties of the objects in your field of vision—their shape, color, and so on. This pathway runs down and forward through the brain's temporal lobe, following the lower arrow in the figure. The Where pathway, by contrast, runs from primary visual cortex up and forward through the *parietal cortex* (the upper arrow in the figure). It contains spatial maps and identifies where objects are located and what direction they're moving in.

For instance, suppose you're driving along, minding your own business, when a yellow trucker hat blows from right to left across the road in front of you. Your eyes are open. Here, roughly, is how you see the yellow trucker hat. First, photons from the sun or your headlights bounce off the yellow trucker hat and into your eye, where they hit specialized neurons in your retina—rods and cones—that detect light. These neurons send a signal through several way stations all the way to the back of your brain—to the primary visual cortex. Here, there are neurons specially tuned to recognize lines in specific orientations. If the hat is oriented upright, neurons that recognize a horizontal line that defines the base of the hat and a nearly parallel line that defines the top of the brim become active. The signal is then sent forward along the What pathway, where neurons dedicated to color detection identify the hat's color as yellow, and neurons dedicated to identifying objects identify it as a trucker hat. Meanwhile, in parallel, the Where pathway is also ruminating on the output from the primary visual cortex, and neurons dedicated to recognizing when objects are near the ground are firing away, as are neurons dedicated to recognizing when objects are moving from right to left.

In a massively simplified nutshell, that's how you use your brain when you are actually seeing. What about when you're merely understanding language about things that could be seen—do you use these same pathways of your visual system to simulate those things? Do you understand language about objects using your What pathway? Do you use your Where pathway to understand language about motion?

Don't Think of an Elephant

A book came out a couple years ago called *Don't Think of an Elephant.*[2] It turns out, not thinking about something is a pretty hard entreaty to comply with. Go ahead and try to not think of an elephant. Don't think of one. Really, try it. If you're like most people, you can't help but see a large, tusked pachyderm in your mind's eye. No matter how heroic our attempt, no matter how steely our resolve, we always succumb to a mental image of the big-eared mammal. It's almost as if you just can't hear the name for something and not create a visual representation of it.

At least, that's what it seems like intuitively. But we should be wary of relying entirely on intuition. People have intuitions about all kinds of things that turn out to be wrong, like your intuition about how a lock works (whatever you think, it doesn't actually work that way) or whether you're sober enough to drive (you probably aren't). What we'd want to see is empirical evidence confirming the intuition that merely mentioning an object evokes a visual simulation. Cognitive psychologist Rolf Zwaan conducted a series of elegant experiments to ask whether this is the case.[3] He started with the simple observation that when objects are in different orientations, they look different. Consider a nail, for example. When it's being hammered into a floor, it's pointed downward, and that looks different from a nail being hammered into a wall. So Zwaan asked whether people automatically construct mental images of objects like nails in a specific spatial orientation when a sentence implies that they are in one orientation or the other.

In their experiment, Zwaan and his students had people first read a sentence like *The carpenter hammered the nail into the floor* or *The carpenter hammered the nail into the wall*. Immediately after reading each sentence, a picture of an object—for instance a nail or an elephant—popped up on the screen. Their task was to decide as quickly as possible whether the depicted object had been mentioned in the preceding sentence. Sometimes the object had not been mentioned (the elephant), but in all of the cases of interest, it had indeed been mentioned (the nail), so the correct answer was "yes." The key manipulation was that when the picture had been mentioned, it could either <u>match</u> the implied orientation of the object or <u>mismatch</u> it. That is, after reading about a nail that was being hammered into the floor, a participant might see a matching picture showing a nail pointing downward or a mismatching picture of nail pointing sideways. And, conversely, after reading about nail being hammered into the wall, they might see a matching picture of a nail pointing sideways or a mismatching picture of a nail pointing downwards. What Zwaan expected was exactly what he found. People were faster to answer correctly when the orientation of the object implied by the sentence matched the orientation of the picture.

Why would people be able to respond faster when the implied orientation of the object mentioned in the sentence matched its orientation in the picture? A very reasonable explanation is that when people read sentences, they construct visually detailed simulations of the objects that are mentioned. When the picture they subsequently see matches the visual details of their embodied simulation—in this case, the orientation of the object—they are able to respond faster

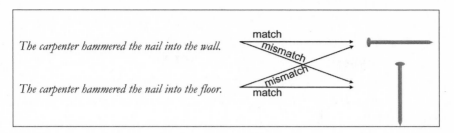

The carpenter hammered the nail into the wall.

match
mismatch

The carpenter hammered the nail into the floor.

mismatch
match

FIGURE 10

because what they see and what they're mentally simulating are more similar. But when the visual details (like the orientation) of their embodied simulation and the picture they see are different, it takes longer to determine that they are the same object. There have now been a dozen or so studies that use variants of this paradigm, confirming our original intuition that people construct visual simulations of the objects they read or hear about.

But how do the details get into our embodied simulations? How do we know to simulate the nail pointing down or horizontally? What's interesting is that there isn't necessarily anything explicit in the sentences that tells you. The orientation of the nail in the sentences above isn't determined by the word *nail*—that's held constant between the sentences. But it also isn't determined independently by the words *floor* or *wall* or even the association between either of these words and *nail*. To prove this to yourself, think about what orientation that the nail has in *The carpenter hammered the nail into the floor* and *The carpenter put the nail on the floor*. In the first, the nail is most likely pointing downward. But in the second, it's most likely horizontal, or if you're feeling malicious, it might be pointing upward. But the point is that the embodied simulation you produce in response to sentences like these is the product of a number of parts of the sentence; in this case at least two nouns (*nail* and *floor*), a verb (*hammered* or *put*), and a preposition (*into* or *on*). You can only figure out which way the nail is pointing by generating a pretty rich model of what's being described, using all these pieces of language and what you know about how you interact with nails, floors, and walls. In other words, the only way that you could know what direction a nail would be pointing in, in a given context, is by accessing your expansive world knowledge about how the different things mentioned in the sentence interact.

And of course the orientation of objects isn't the only visual detail that you mentally simulate when you read or hear sentences. When a sentence implies that an object has a particular shape, you mentally represent it having that shape. For instance, *The ranger saw the eagle in the nest* might imply an eagle with folded wings. By contrast, when we hear a sentence like *The ranger saw an eagle in the sky*, there's good reason to believe that the eagle has outstretched wings (physics and all that). Zwaan and his students performed another experiment, in which they exchanged the manipulation of object orientation for a manipulation of object shape.[4] Everything else was the same as in the *nail* experiment—each sentence mentioned an object with an implied shape, and people had to read the sentence and then decide whether an object had been mentioned in the sentence. And this time, just like in the orientation study, people were faster to press a button indicating that the picture had been mentioned in the preceding sentence when the implied shape of the object matched the shape of the image. In other words, when a sentence implies that an object has a particular shape, we construct a mental representation of the object with that implied shape—just as we represent the implied orientation of objects.

The results from these two studies seem to imply that people processing sentences automatically construct perceptually detailed embodied simulations of described objects. But before taking that to the bank, we need to consider a reasonable objection to this methodology. Imagine you're a participant in the experiment. You know that your task is to read sentences and then decide whether a subsequently displayed picture depicts an object mentioned in the sentence. Because you know this, you might consciously build mental images of the objects you read about. And this might well be quite different from what you normally do when using language in the real world. That is, the

The ranger saw the eagle in the nest.

The ranger saw the eagle in the sky.

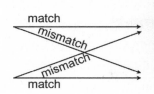

match

mismatch

mismatch

match

FIGURE 11

effect might be due to demands of the task that are different from the demands placed on you when normally using language.

Zwaan and his students addressed this possibility by conducting a follow-up experiment, in which they used the same shape-manipulating sentences (such as the *eagle* sentences) in a slightly modified task.[5] Instead of deciding whether a depicted object had been mentioned in the preceding sentence, participants simply named each object as they saw it. This meant that—for the participants—there was no longer any obvious relation between the sentences and the pictures, and it wouldn't help them at all to construct detailed embodied simulations of the objects mentioned in the sentence to perform the task. Nonetheless, the results were the same—participants were faster to name a picture when it matched the shape of the preceding sentence. This shows us that people mentally simulate the visual details of objects, even when the experimental task does not lead them to. It appears to be quite natural for people understanding language to produce embodied simulations of the things that they read or hear about.

Orientation and shape aren't the only properties of objects that could potentially appear in our embodied simulations. We perceive objects in the real world in large part through their color. Are the embodied simulations we construct while understanding language in black and white, or are they in color? It seems like the answer should be obvious. When you imagine a yellow trucker hat, you feel the sub-jective experience of yellowness that looks a lot like yellow as you would perceive it in the real world. But it turns out that color is actu-ally a comparatively fickle visual property of both perceived and imag-ined objects. Children can't use color to identify objects until about a year of age, much later than they can use shape.[6] And even once they acquire this ability, as adults, people's memory for color is substan-tially less accurate than their memory for shape,[7] and they have to pay closer attention to detect changes in the color of objects than in their shape or location.[8]

And yet, with all this going against it, color still seeps into our em-bodied simulations, at least briefly. One study looking at color used the same sentence-picture matching method we've been talking about.[9] People read sentences that implied particular colors for objects. For instance, *John looked at the steak on his plate* implies a cooked and therefore appropriately brown steak, while *John looked at the steak in the butcher's window* implies an uncooked and therefore red steak. In the key trials, participants then saw a picture of the same object, which could either match or mismatch the color implied by the sentence— that is, the steak could be red or brown. Once again, this method pro-duced an interaction. Curiously, though, the result was slower reactions to matching-color images (unlike the faster reactions to matching shape and orientation images in the previous studies). One explanation for why this effect appears in the opposite direction is that perhaps people processing sentences only mentally simulate color briefly and then sup-press color to represent shape and orientation. This might lead to slower responses to a matching color when an image is subsequently presented. (It's important to note that this reversed effect doesn't mean that people aren't simulating color. That's an inference you might be ex-pected to draw if there were a *null effect*—if there were <u>no</u> difference in responses times to the differently colored images, regardless of which sentence they followed.) These complexities notwithstanding, this study indicates that when a sentence mentions an object, people mentally simulate the color of that object—if transiently.

All these studies point to the same conclusion. Hearing or reading language about objects leads people to mentally simulate those objects. So to come back to the issue we started with, it's not your fault if you can't help seeing an elephant when told not to. It's just an automatic part of understanding language.

The Elephant in the Room

The visual properties we've been looking at, like an object's orientation, shape, and color, are all properties that brain calculates in the What pathway. But we oftentimes describe objects as being located in particular places or moving in particular directions. If someone tells you that *there's an elephant in the room*, do you simulate not only what the elephant looks like, but also where it might be? The broader question here is whether people mentally simulate location and motion—features that are computed using the Where pathway.

Let's start with the simplest case. Objects tends to appear in specific locations. Grass is typically below your feet and the ceiling is usually above your head. And as a result, you know when you hear words for such things where you should expect to find them. A puddle, a cloud, and a manhole cover all have canonical locations (down, up, down). So, we might begin by asking, does language about objects like these engage the Where pathway of your visual system?

We ran a study in my lab that investigated this question, based on the Perky effect discussed in the last chapter. (You will recall that in her work, Perky found that people have a harder time perceiving objects when they are simultaneously visually simulating objects in the same location.) In our study, we presented participants with simple recorded sentences, like *The sky darkened*, or *The grass glistened*.[10] In all these sentences, the noun (like *sky* or *grass*) was an object with a canonical location, and the verb (like *darkened* or *glistened*) didn't imply anything about the location or motion of the object. Very soon after hearing each sentence (just one fifth of a second), the participants saw a shape on the computer screen, which was either a circle or a square. They had to decide whether they saw a circle or square and push an appropriate button to indicate their choice. The shape itself also appeared for only a fifth of a second, which made the task quite difficult. Although the participants thought we were interested in circles and squares, we were really manipulating another feature of the shapes, which was where they appeared on the screen. For the sentences we were interested in, the shapes always appeared either in the upper part of the screen or in the lower part of the screen, always horizontally centered. What we wanted to see was whether processing a sentence about an object that is canonically found in an up location (like *sky* or *cloud*) would affect how long it took people to perceive a shape appearing in the upper part of the screen, and vice versa for objects that canonically appear in a lower location (like *grass* or *puddle*).

What we found was surprising. It took significantly longer—about 30 milliseconds more—for people to categorize the circle or square when it appeared in the same location where a just-mentioned object would canonically appear. You might think this sounds counterintuitive. Why would it take longer to say that an object was a circle or square when you saw it in the upper part of the screen after hearing a sentence describing an object in the upper part of the visual field? Shouldn't this be faster because the objects are in the same location? Actually, no. Recall that when people perform visual simulation, they use the same parts of their brain that are responsible for actual vision. When you mentally simulate objects that would normally appear above you, you use the parts of your vision system that are dedicated to perceiving objects in that location. And you just can't do two different things with that same part of your brain at the same time—it's relatively hard for you to use the same neurons that represent a particular region in space to both simulate a cloud and also perceive a circle or square. This study shows that the Perky effect is alive and well not only when people are told explicitly to mentally imagine objects, like in her original work, but also when they simply hear sentences about objects. It also demonstrates that the embodied simulations people activate while listening to language represent objects in their expected locations.

Now, you might object that this is all fine and good for objects that have a canonical location. But many objects move around. It's not clear, for example, where a *donkey* usually is, or for that matter a *glass*, a *stone*, or a *pipe*. But when you use these nouns in sentences with directional verbs like *climb* and *drop*, you can infer that they end up in a specific location. That leads us to ask whether a sentence like *the glass dropped* leads you to mentally simulate the object in the lower part of your imagined visual field. And likewise, does *the donkey climbed* cause you to perform mental imagery in the upper part of your visual field?

We conducted another experiment to address this question.[11] We again presented participants with spoken sentences, but this time the nouns (like *donkey* or *glass*) were neutral for upness or downness, and it was the verbs (like *drop* and *climb*) that indicated a location. We expected that if the verbs in these sentences could cause people to mentally simulate the locations of these objects—even though the objects didn't have canonical locations—we should see the same Perky effect. That is, we expected it to take people longer to categorize a circle or square when it appeared in the same location as the object mentioned in the preceding sentence. And again, this is what we found, with significantly delayed responses (about 40 milliseconds slower) when the sentence described an object in the same location as the shape to be categorized. This tells us that when you hear about an object that doesn't have a canonical location, you still perform an embodied simulation of its location, which you construct dynamically by putting together pieces of the sentence.

Some of the most exciting evidence about how language leads you to visually simulate comes from eye-tracking studies. Eye tracking is a data collection technique that uses high-speed cameras and specialized data analysis software to monitor where exactly a person is looking while performing some task.[12] One of the main benefits of using eye movements as a way to measure what's going on inside people's heads is that, despite what you might think, most eye movements are not under conscious control. In fact, our intuitions about how our eyes move are often quite inaccurate. For instance, how do you think your eyes move as you read the words on this page? Many people believe that their eyes move continuously along each line of text. In fact, though, it was discovered in the eighteenth century that while reading (and doing most everything else) our eyes move in leaps (or *saccades*). And they leap a lot. During most of our waking lives, our eyes saccade from one fixation point to the next dozens of times a second, depending on the interesting things around us that we're paying attention to. And not only are these movements not under our conscious control, but they couldn't be. It's just too many movements that are too fine in too little time. As a result, because eye movement is mostly unintentional and because it is so closely linked to attention, tracking eye movements gives us a pretty good idea of what the visual system is up to. The obvious question that eye tracking can answer with respect to language is whether the visual system is doing similar things when you're understanding spatial language and when you're perceiving actual things in the real world.

Cognitive scientist Michael Spivey was the first to address this question, using an ingeniously deceptive setup.[13] The issue that Spivey had to overcome is that, although natural eye movements are mostly not under conscious control, people are often cautious when they know their eyes are being tracked. So in order to ensure that participants didn't guess the purpose of their experiment or modify their eye movements in conscious ways, he had them take part in a sham experiment in which they had to follow instructions about moving objects around on a table while their gaze was being tracked using an eye tracker. Then, during a "break" in this "experiment," participants rotated their chair so they faced a white screen. They were told that the eye tracker was being turned off and that they would hear a couple of short stories. Unbeknownst to the participants, this was when the real experiment began.[14] As they heard the stories, their eye movements were in fact recorded. Meanwhile, the stories described scenes that, if you were to actually view them, would involve scanning in one direction or another. Compared with a control story, which involved no movement, there were stories that would entail upward, downward, rightward, or leftward eye movement. For instance, contrast the upward and downward stories below.

Upward story:
Imagine that you are standing across the street from a forty-story apartment building. At the bottom there is a doorman in blue. On the tenth floor, a woman is hanging her laundry out the window. On the twenty-ninth floor, two kids are sitting on the fire escape smoking cigarettes. On the very top floor, two people are screaming.

Downward story:
Imagine you are standing at the top of a canyon. Several people are preparing to rappel down the far canyon wall across from you. The first person descends ten feet before she is brought back to the wall. She jumps again and falls twelve feet. She jumps another fifteen feet. And the last jump, of eight feet, takes her to the canyon floor.

Spivey hypothesized that if people listening to these narratives were mentally simulating these scenes as they processed the stories, they should move their eyes in the direction of the described scene. That is, their eyes should make more upward movements while listening to the upward story condition and more downward movements during the downward story condition. This is precisely what they found. People's eyes jumped significantly more in the direction of the movement in the story than in other directions. The best explanation of these findings is that when processing stories about events taking place in particular locations, understanders construct embodied simulations of objects in those same locations, and their eyes are dragged along for the ride.

Subsequent work by a group of researchers in Sweden has extended this finding.[15] They tracked eye movements while people looked at pictures of objects on a landscape. Then they took the picture away and asked these participants to describe what they had seen while continuing to track their eyes. They also tracked their eye movements while they listened to other people's scene descriptions. What they found was that when people heard or talked about specific parts of the picture, their eyes moved to the places on the now blank screen where those objects had originally been. And if that wasn't enough, they even conducted the retelling and listening parts of the experiment in complete darkness and once again found that they still showed the same eye movement patterns.

But what about when objects are described as being in motion? Do we construct dynamic simulations in which we virtually see objects moving through space? This question was first addressed by Daniel Richardson and his colleagues, who conducted a Perky effect experiment, appropriately also at Cornell, much like the ones described above.[16] People listened to sentences that described either vertical motion (like *The strongman lifts the barbell*) or horizontal motion (*The miner pushes the cart*). Then they saw a shape that was either a circle or square and had to press the appropriate button as quickly as possible, just like we saw before. In this experiment, though, the shape could appear not only directly above or below the center of the screen, but also immediately right or left of center. The experimenters reasoned that if understanders were constructing mental representations of the entire path of motion—for example from high to low versus left to right—then this should slow down their perception of shapes when they occurred anywhere along the same axis (up-down or left-right). So *The strongman lifts the barbell* should produce slower responses to shapes appearing either in the upper or the lower parts of the screen but faster responses to shapes appearing on the left or right of the screen. Conversely, *The miner pushes the cart* should produce slower responses to shapes appearing along the same axis—when they're to the right or left of center but faster responses when they're above or below the center of the screen. This was just what they found.

Although this result might suggest that people understanding sentences mentally represent the trajectories of described motion, there could be other explanations. For instance, it could be that hearing a sentence about upward motion causes you to mentally represent a static simulation of the trajectory—something like the motion line you would see in a cartoon to indicate motion. To test whether the visual simulations underlying language understanding are really dynamic—whether you really represent motion in your head—Rolf Zwaan and his colleagues introduced a very clever method.[17] They had people listen to sentences that described the motion of a ball toward or away from the observer (e.g., *The pitcher hurled the softball to you* versus *You hurled the softball to the pitcher*). After each sentence, they saw two sequentially presented images, each of which was on the screen for half a second. The two pictures were of either the same object (like a softball) or different objects (like a softball and a watermelon). The participants had to decide whether the two pictures depicted the same object or not. But the experimenters were not at all interested in the cases where there were two different pictures. The real manipulation dealt only with the matching pictures. What the experimenters did was to slightly modify the size of the second of these pictures, so that it was slightly larger or slightly smaller than the first picture. Because the two pictures were shown in quick succession, their difference in size produced an illusion of motion either toward the viewer (if the second picture was larger) or away from the viewer (if it was smaller). The researchers reasoned that if people perform dynamic embodied simulations of motion driven by sentences describing motion toward or away from them, then these simulations should yield faster categorizations of objects that appeared to be moving in the same direction. And this is what they found; although small (only 21 milliseconds on average), the difference in response times was significant and in the predicted direction. People produced responses more rapidly when the illusory movement of the balls matched the movement described in the sentence.

Our use of the Where pathway during language understanding ensures that when we think of an elephant, we simulate not only what it looks like but where it is and where it's going. More broadly, it appears that understanding meaning, at least in part, happens in the mind's eye.

Almost Like Being There

These visual experiences that people have while understanding words—how exactly do they fit together? One can imagine one of several different possibilities. First, it could be that the things we mentally simulate—nails, steaks, or elephants—we see from a God's-eye view. This is plausible because we know that there are neurons in the vision system that are sensitive to specific objects not from any particular viewpoint but independent of viewpoint (so-called viewpoint-invariant neurons).[18] Maybe these are the parts of the vision system we use when understanding language. Maybe the entirety of the nail that we see is visible in our mind's eye, not just the side facing us. A second possibility is that perhaps the embodied simulations that we construct actually do adopt a particular viewpoint but one that's insensitive to the details of the particular sentence. It's well known that people often activate mental representations of objects from so-called canonical perspectives, and perhaps the embodied simulations people activate are basically concatenations of sequences of objects, each viewed from its canonical perspective. Or, finally, there's a more radical possibility. What if understanding language were in some way akin to actually being there, to experiencing the events that the language describes? In this case, you should again view objects and events from a particular perspective but one that's dependent on the details of the sentence and its context. This idea is sometimes known as the *immersed experiencer* view.

How can we distinguish among these possibilities? You might think that the immersed experiencer view holds more merit, and, if so, you're not alone. This is in fact the view shared, often implicitly, by a number of researchers.[19] But the thing is that all three possibilities

(God's-eye view, canonical view, and immersed experiencer view) are consistent with basically all the research we've reviewed thus far. So whichever one we might think sounds more reasonable, we need a way to distinguish among them empirically. And the way to do this is by looking at differences in the predictions they make about new observations we might be able to collect.

The first view to take by the horns is the idea that people adopt a viewpoint-invariant God's-eye view in their embodied simulations. In principle, this is possible, but there's actually a lot of evidence that words trigger embodied simulations of objects from a specific vantage point—evidence against viewpoint invariance. Consider the elephant you imagined earlier when told not to. You surely didn't view it from just any arbitrary perspective, like from below, from behind, or above. You almost certainly viewed it either from the side or head-on, or possibly from some intermediate position between the two. In addition to a preferred angle, you probably viewed the elephant from a preferred distance. Not so close that you could see the individual hairs on its skin, not so far that it would be indistinguishable from a pile of tires. Instead you probably saw it from a distance where you could just see the whole elephant. Angle and distance make up the perspective from which a real object is viewed or a described object is simulated. There's a good deal of evidence that people tend to adopt one perspective over others when conjuring up embodied simulations in response to words. Early work had participants describe the consciously accessible mental images they produced for given words.[20] Unsurprisingly, people reported—like you undoubtedly would—that their mental images tended to take on one particular perspective, which has come in the literature to be called the *canonical perspective* for an object, just like your mental image of an elephant. But self-reports like these are, as we know, finicky and unreliable—it's preferable not to directly ask people to reflect on what's going on in their own minds. Fortunately, the finding that people prefer to mentally represent objects from certain canonical perspectives has been frequently corroborated by further work using other, more objective measures.[21]

It's worth taking a little detour here to talk more about canonical perspectives, because they're actually quite interesting in and of themselves. When we look in a little more detail at the work on canonical perspectives, it's clear that although words for objects—like *elephant*, *teacup*, and *football helmet*—evoke mental representations in which they are seen from particular perspectives, it's quite hard to figure out why people use the particular canonical perspectives they do for specific objects. The problem is that there are a number of potential factors. You might be more likely to adopt a perspective that's more typical.[22] For instance, you probably interact more frequently with elephants from the side or front than from below or the back (if you're lucky). Another factor is what perspective is more informative.[23] If you imagine a teacup from certain angles, you might not be able to view the handle or determine what shape it is. So you might prefer to adopt a perspective in which the handle is in view, because this might be important information about the cup. Although there are debates over exactly what leads people to adopt specific canonical viewpoints and what causes variation between and across individuals, it is implicit in all this work that hearing or reading words for objects drives people to mentally simulate those objects from a particular visual perspective. All of this is evidence against viewpoint-invariant, God's-eye mental representations of objects.

So that leaves us with two possibilities remaining. We know that when people hear or read words, they mentally simulate the pertinent objects from a given perspective. Over the course of understanding a longer, more connected piece of language, like a sentence, paragraph, or even a chapter or book, do people immerse themselves into the scenes, viewing objects from a perspective such that the way the objects look in their mind's eye are like how they would appear to someone experiencing them firsthand? Or are embodied simulations to longer discourse just a smattering of canonical images, popping up one after the other? Fortunately, these two views make different, testable predictions.

Here's one. The immersed experiencer view claims that when you're understanding language, you simulate what it would be like to experience the scene that's described. If the immersed experiencer view is right, then we should observe that people adopt different perspectives when they process language in which the person experiencing the scene would have a different viewpoint. For instance, this view predicts that you will adopt a different perspective on an elephant when processing a sentence about *grabbing an elephant by the tail* versus *sitting on an elephant's back*. And these different perspectives should produce measurably different embodied simulations. By contrast, if people are just activating canonical views of objects, the sentence context shouldn't affect the particular perspectives that people adopt when simulating objects.

Italian psychologist Anna Borghi and her colleagues investigated this issue by having people read sentences about objects that implied a perspective that was either internal or external to the object.[24] For instance, *You are driving a car* implies that you are in the car, but *You are washing a car* implies that you are outside of it. After they read the sentence, the participants saw an expression identifying an interior part of the car, like *steering wheel* or *gas pedal*, or an exterior part of the car, like *tires* or *antenna*. And they had to decide as quickly as possible whether the expression described a part of the previously mentioned object: for instance, whether a steering wheel or antenna was a part of a car. The results showed that participants responded faster when the perspective suggested by the sentence matched the location of the part. In other words, first reading the sentence *You are driving a car* made people faster to decide that a steering wheel was part of a car but slower to decide that tires were part of a car. And vice versa for *You are washing a car*. The conclusion? It appears that language does manipulate what perspective you adopt when you mentally simulate objects. This implies more generally that people reading sentences project themselves into mentally simulated experiences of the described scenes.

There is a second testable implication of the immersed experiencer view. When you see an event in the real world, you use your visual system, which, with all its limitations, does not always render the

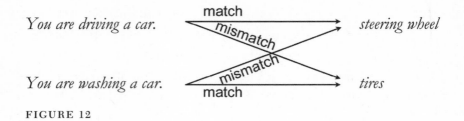

FIGURE 12

scene perfectly. All sorts of factors (rain, fogged-up glasses, blurred vision) can obscure and decrease the visibility of otherwise observable things. If we really do immerse ourselves in a virtual simulation of experiencing a described event, then we should represent described objects as measurably more or less visible. And how visible an object is should depend on how visible it would be to someone experiencing the actual scene. Researchers at Florida State tested this prediction in a clever way.[25] They presented participants with sentences about people viewing objects through an occlusive or a transparent medium. For instance, contrast *Through the clean goggles, the skier could easily identify the moose* with *Through the fogged goggles, the skier could hardly identify the moose.* Participants then saw a picture of an object, and in the critical trials, it was the object mentioned in the preceding sentence (a moose, in this case). They had to decide whether that object had been mentioned in the sentence. The trick was that the moose could be depicted in high or low resolution, as seen in Figure 13. The experimenters found that, surprisingly, people were faster to identify the matching picture—high resolution pictures after sentences about clear vision and low resolution pictures after sentences about obscured vision. This is compelling evidence that people are simulating what it would be like to "be there" in the scene because, from any other viewpoint than inside the skier's goggles, the moose wouldn't be harder to see when the goggles were fogged up. And yet people act like they were having more trouble seeing the moose when simply hearing about someone else having trouble seeing the moose.

Through the clean goggles, the skier could easily identify the moose.

Through the fogged goggles, the skier could hardly identify the moose.

match

mismatch

mismatch

match

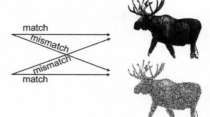

FIGURE 13

Other experimental evidence pointing in the same direction comes from another recent study, in which people read stories that, similarly to the high and low visibility sentences, described a person whose view of a number of objects was either clear or obscured.[26] After the story was over, they were asked to say whether objects had been mentioned in the story, and they were significantly faster to do so when the objects had been described as clearly visible than when they had been described as obscured. Again, it seems that when we hear or read about objects, we mentally simulate them from the perspective of someone actually experiencing the scene—not from a God's-eye view and not from a canonical viewpoint.

There's even some recent work showing that people mentally simulate an object differently according to the distance a sentence suggests they would be from it. This is work that a graduate student, Bodo Winter, did with me.[27] Perspective involves not just the angle that you see an object from and its ease of visibility but also its distance away from you. Compared with objects viewed from close up, faraway objects look smaller and blurrier (more so depending on your age). So if the immersed experiencer view is correct, then when a sentence indicates that an object would be seen from farther away, you should see it in your mind's eye as smaller and blurrier than when it's described as close by. We tested this in much the same way as several of the previously discussed studies. People read sentences like the ones in Figure 14, which suggested that the object (in this case a golf ball, but there were others, like axes and sheep), was relatively close

You are staring at the golf ball in your hand.

You are staring at the golf ball in the sky.

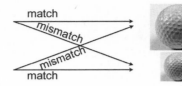

FIGURE 14

or relatively far from the observer. Then people saw an image that either depicted the object just mentioned or not. We were only interested in the cases where the object had indeed been mentioned, in which we presented either a large and clear image of the object or a small and blurry one. Like previous work, we found that people were faster to verify the object in the matching conditions than the mismatching ones; responses were about 50 milliseconds faster to a small image after a sentence about the object being far, and about 50 milliseconds faster to a large image after a sentence about the object being close. So when you understand language about objects that are implied to be closer or farther, you represent those objects as having visual characteristics appropriate to their distance.

Rolf Zwaan put it eloquently when he first described the immersed experiencer idea: "Comprehension is the vicarious experience of the described events through the integration and sequencing of traces from actual experience cued by the linguistic input."[28]

As it turns out, people don't just visually simulate from a given perspective. The next chapter reveals that what things look like in your mind's eye is only the beginning of the story.

CHAPTER 4

Over the Top

If you're like me, your victories in arm wrestling are few and far between, except against the young, the infirm, or the unconscious. This is at least in part because where most people have normal, human-sized forearms, I have popsicle sticks. But if you're like me, and if this failing causes you mental anguish, fear not. It turns out that being built like Popeye is not the only way to get the upper hand—as it were—in arm wrestling. No, like in other sports, technique is nearly as important as physique. Case in point; consider the compelling argument for technique made by Sylvester Stallone in *Over the Top*, undisputedly the greatest arm-wrestling film in the history of world cinema. In the film, Stallone plays a truck driver named Lincoln Hawk, who enters an arm-wrestling competition with the noble objective of winning back the respect of the son he abandoned. Hawk's road to victory is predictably long and full of apparently insurmountable setbacks, but he succeeds because he has a secret weapon—the "toproll" technique. If you find yourself needing to make amends with an estranged offspring through arm wrestling, here's how you can do it using moves like Sly.

First, scoot your elbow forward, toward your opponent. This makes your forearm stand more upright, and, as a result, your hand will be higher than your opponent's, giving you a leverage advantage. Then as you grip your opponent's hand, grab it higher than you normally would—instead of hooking your thumb around the base of theirs, slide your hand up a little so you're grasping farther up toward the tip of their thumb. When you start the actual wrestling, curl your wrist toward you, using the leverage that you have over your opponent's hand, which will flex your opponent's hand backward and give you an additional leverage advantage. From there, you should be able to topple even a stronger opponent's hand toward your side. And of course, if life imitates art, by doing so, you will also topple any barriers between you and your estranged children.

So that's a little inside insight on arm-wrestling technique. But let's pause for a moment to reflect on what you've just done. You read a couple hundred words, and, from those words, you have created new knowledge, knowledge that you could put into action if you wanted to. If you take a step back, this is a pretty amazing feat. No other species has anything like the capacity that you have to look at lines and curves and dots on a page, interpret them as describing actions you've never performed or possibly even seen before, and translate them into plans for action that you can execute using your muscles. Perhaps this doesn't seem like such a remarkable capacity to you; after all, you do it all the time, when you read about how to contort your body into a new yoga pose, the "upside-down lotus," or when a company memo reminds you about the best way to lift boxes of books without straining your back. But one way to achieve the appropriate sense of wonder is to ask yourself, if you had to engineer a biological system that could do what you can do—translate words into actions—how would you do it?

It's an important question, because in fact, a lot of what we say and write has to do with people moving around in the world. Sometimes it's in the form of instructions—when to press down on the clutch and how quickly to release the gas, that you should stop scratching that scab or it will scar, or how far you have to roll up your sleeves to keep them clean while you're hollowing out a jack-o'-lantern. People also give descriptions of things they have done or witnessed—how they tripped over a skateboard in the garage or how, in this game against the Lakers, Dr. J reached from completely out of bounds under the basket to scoop in a layup.

How do we take words and sentences that we hear or read and convert them into thoughts about actions that we might never have performed or seen and, more interestingly, into instructions for how we might perform those same actions with our own bodies? Any description of action by necessity leaves out lots of details; how do we make inferences about how our bodies would move were we to perform these actions? For instance, in reading the toproll description, you might have inferred that when you push your elbow toward your opponent, you nevertheless keep it down on the table; and that when you do this, you need to lean in a bit, because otherwise your arm will be too far from your body to generate much power. Or you might have inferred that as you roll your hand over your opponent's, you twist your hand so that your palm faces down. How do you know all this even though it was never mentioned?

The short story, as I'm sure you will have no difficulty guessing by this point, is that people understand language about physical actions by performing motor simulations of the described actions, using the brain systems responsible for coordinating and executing actual action. But first, we need to know a little more about how the brain represents actions. And for that, we turn to monkeys.

Monkey Do, Monkey See

In the 1980s and 1990s, neuroscientists at the University of Parma, in Italy, were studying exactly how neurons in the brains of macaque monkeys control their body movements. To do this, they inserted electrodes into the monkeys' brains, targeting individual neurons in motor regions, and measured the electrical activity that each neuron produced when the monkey performed different actions. The first thing they found was fascinating but, in retrospect, not particularly surprising. The monkey brain contains within it neurons that only fire to control particular actions. So there are neurons that selectively drive what's called a *precision grasp*—pressing the thumb against the fingertips, as you might to pick up a paper clip—and different neurons that fire for a *power grasp*, the way you might grasp the handle of a hammer (or the hand of your arm-wrestling opponent). I say that this should be unsurprising simply because we know that monkeys are able to select from an array of different actions, and, because the motor system is in control of these actions, it must have distinct ways to represent each one.

The more remarkable and unexpected finding—one that some neuroscientists have called the most important in decades[1]—was that some of those very same neurons, dedicated to specific actions on specific objects, became active not just when the monkey was performing an action but also when it was observing someone else perform the same action.[2] In other words, if one of the scientists were to pick up a paper clip, the paper clip–grasping neurons in the part of the monkey's brain that controls action would start to get excited. But the hammer-grabbing neurons would not. And vice versa if the monkey saw the experimenter pick up a hammer. This is depicted in Figure 15.

Along the top are the conditions—an experimenter picking up a piece of food, the monkey picking up the piece of food, or, to make sure the neurons aren't just responding to someone interacting with the food, an experimenter grasping the food with pliers. Below these pictures, you see a representation of each time the neuron in question became excited enough to fire (these are the small vertical lines that look like this: || || ||||| |||||). The x-axis is time, so you can see from the beginning of the action (the round dot), when the neuron fired. And at the very bottom is a histogram that summarizes the firing of this same neuron over the repeated trials depicted above. You can see that this particular mirror neuron fires most concentratedly when the monkey is grasping the food, to a lesser but still substantial degree when the experimenter grasps it, and almost not at all when the experimenter uses pliers on the food.

The discovery of these *mirror neurons*—so called because they seem to encode both the performance and observation of specific ac-

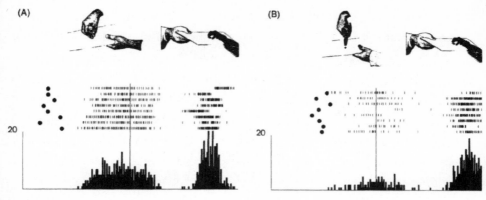

FIGURE 15

tions—is compelling for several reasons. Foremost among these is that these "mirror" brain structures offer an exciting potential explanation for how it's possible for our macaque cousins, and for us, to recognize, predict, and understand the physical actions of others. If recognizing another's actions involves activating some of the same brain structures that we use to perform the same actions, then we can use all the neural machinery we have that's dedicated to action (machinery for predicting outcomes of actions, for calculating the goals of actions, etc.) to reason and make inferences about the actions we see, just as we would about actions we're enacting ourselves.

Most cognitive neuroscientists now believe that the same sort of representational architecture is at play in humans. But recording from individual neurons is a stickier enterprise in humans than in monkeys, because we have to actually open up a hole in the skull to insert the electrode, and the institutional boards that regulate research with humans subjects tend not to look too favorably on this practice. So we can't just take single-cell recordings of mirror neurons in the human brain. However, in humans, we can look at populations of neurons through brain imaging. And, indeed, brain imaging research shows overlaps in brain activation while people perform and perceive specific actions.[3] That is, although there isn't direct evidence of individual mirror neurons in humans, there is evidence of mirror systems that appear to behave as one would expect if indeed human motor control systems doubled as motor action perception systems.

Mirror neurons, and the mirror systems they're a part of, are relevant to the question of language in the following way. Perceiving someone else perform an action results in the activation of your own motor system. What if understanding language about action hooks into this same system, such that the motor system can be used not only when people are acting, perceiving action, and thinking about action, but also when they're understanding language about acting?

The Give and Take of Motor Simulation

The simplest way to test whether language about actions engages the parts of the brain responsible for performing those same actions is to have people perform those two tasks at once and see if they interact. The line of reasoning here is basically the same one that motivated the experiments on visual imagery of shape and orientation we saw in the last chapter. If you have someone read or listen to a sentence that describes a motor action of a particular type—say, someone moving their hand away from their body—and then have them perform either that very same action or an incompatible one (like moving their hand toward their body) then they should be faster to perform the compatible action than the incompatible one.

An illustration is probably in order here. Suppose you're at a computer, with three buttons laid out in front of you, as in Figure 16. What you're supposed to do is to push and hold down the gray button with your dominant hand, which will make a sentence appear on the screen. As long as you hold down the gray button, the sentence will stay there. You read the sentence and then decide whether it makes sense or not. If it does, you use your same hand to push the black button, and, if not, you push the white button. The critical point is that you have to move your arm away from your body in order to push the black button but toward your body for the white one. So your response requires motion in a particular direction, which of course in-

FIGURE 16 The setup of action-sentence compatibility experiments—the participant presses the gray button to see a sentence, and then the black or white button to indicate whether the sentence makes sense or not.

volves a particular set of muscles, which are driven by a particular set of neurons in your motor cortex. Now the interesting design feature of this experiment is that some of the sentences that pop up when you press the gray button describe motion toward the body, and others describe motion away from your body. For instance, you might see a sentence like *You handed the puppy to Katie*, which, were you to perform the described action, would require you to move your hand away from your body. Or you might see a sentence like *Katie handed the puppy to you*, which describes movement toward your body. Then, halfway through the experiment, the locations of the black and white buttons are swapped, so you make responses over the course of the whole experiment that are both compatible and incompatible with sentences about motion away from you and motion toward you.

If understanding sentences about actions activates the parts of the brain responsible for performing those actions, then reading a sentence about a particular action should lead you to respond faster when you're performing a compatible action than an incompatible one. For instance, after you read *You handed a puppy to Katie*, you should be relatively fast to release the gray button and push the black button where it's positioned farther away from you as in Figure 16. But if we switch the black and white buttons so that the black button is now closer to you, you should be slower to press it in response to *You handed a puppy to Katie*, because moving your hand towards your body is incompatible with the puppy-transferring action you're mentally simulating. And the reverse should be true after reading a sentence like *Katie handed a puppy to you*—in this case you should release the gray button and push the black button faster if it's located close to you—compatible with the action "you" perform in the sentence.

There have now been about a dozen studies in this paradigm, called the action-sentence compatibility effect.[4] The original work by Arthur Glenberg and Michael Kaschak at the University of Wisconsin showed that different sorts of sentences produce an action-sentence compatibility effect: declarative sentences that describe hand actions, like *You handed Katie a puppy* or imperative ones that tell you to perform hand actions, like *Grab your nose*.[5] Figure 17 shows what the results look like: when the black button is placed near the participant, responses are faster to sentences describing motion toward "you" than sentences describing motion away from "you." And vice versa when the black button is placed in the position farther away from the participant.

Studies like this have been influential, because they seem to show that people understand language about actions using embodied simulation of those same actions. But there's a concern that has been raised with their design. Namely, it might not be the direction of your motion to press a button per se that's compatible or incompatible with a sentence, but rather the location of the button. That is, this might not be an <u>action</u>-sentence compatibility effect, but rather a <u>location</u>-sentence compatibility effect. If this is true, then experiments like this

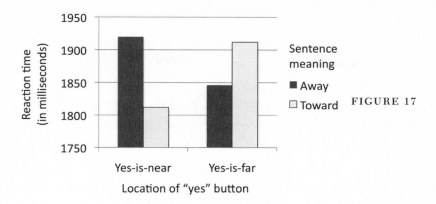

FIGURE 17

don't actually show that people engage mental representations of action while understanding action language, they show that people are engaging mental representations of the locations of the things that are described (like the puppy). And we already know from the experiments we saw earlier on clouds and grass that people do mentally represent the locations of things.

The researchers who first came up with this methodology responded by performing a new experiment. In this experiment, there was no gray button, just a black one and a white one. Participants held one hand above each of the two response buttons, so that they didn't actually have to dynamically move their hand toward or away from their body to press the buttons. If it were just location that was important, then there should be a compatibility effect in this experiment just like in the previous one—sentences about action toward your body should lead you to press a button faster when it was closer to your body, even though your hand was already positioned there. But if it's the action itself—the moving of your hand toward or away from your body—that's the source of the compatibility effect we saw earlier, then there should be no compatibility effect when the hands are positioned directly above the buttons and no forward or backward action is required. And sure enough, the researchers found no effect when the hands were positioned over the buttons. So we can be fairly confident that it's the direction of action and not merely the ending location that's producing the compatibility effect.

But, beyond this concern, you might wonder whether these find-ings really say anything about how people understand language about action in general. After all, all the sentences used in this original study were about actions involving *you*, whether about things *you* have pur-portedly done (like *You handed Katie the puppy*) or things you ought to do (like *Grab your nose*). And in a way, these are the sentences that—at first glance—one would think people would be <u>most</u> likely to perform motor simulation to understand. In principle, it seems less likely that we'd engage our motor system to understand sentences that describe <u>other</u> people performing actions, like *Mary put in her ear-ring* or *Fred offered the tray of hors d'oeuvres to the guests*. So, if we're to believe that people use their motor systems to understand language about action in general, then we'd need to see evidence of an action-sentence compatibility effect for language about other people performing actions, too.

And the short answer is that we do.[6] Experiments conducted in my lab have found that we are faster to move our hand away from our body after reading a sentence about someone else, like Mary or Fred, moving their hand away from their body and faster to move our hand toward our body when the sentence describes someone else moving their hand toward their body. In short, it doesn't appear to matter who is described as performing the action—language about actions in gen-eral results in motor simulation.

The Graspasaurus and Other Magnificent Toys

So people simulate performing the actions that they read or hear about. But so far we've only seen this in a gross way—people simulate the direction they would move their hand and arm in. But the motor system encodes fine detail of the movements we know how to per-form—grabbing paper clips versus hammers, for instance. Are the motor simulations activated by language this detailed? For instance, think again about sentences like *Mary put in her earring* or *Fred of-fered the tray of hors d'oeuvres*. To actually perform these actions, you would need to shape your hand in a particular way. You put in an earring by pinching together your thumb and probably your index finger. But when you offer someone a tray, you might be more likely to hold your palms open and facing upward. When you understand language about these actions, does your motor system simulate the actions all the way down to the last minute implied detail, like handshape?

We've done some research on this question in my lab. In one study, we had people read sentences describing actions they would perform using either a flat palm handshape (*Paul carried the watermelon*) or a closed fist shape (*Sue carried the marble*). The participants then had to perform either a compatible or an incompatible action—logically just like in the action-sentence compatibility effect studies described earlier. So in responding to the sentences, participants pressed a large button with either their open palm (compatible with carrying a watermelon) or their closed fist (compatible with carrying a marble). What we found was an action-sentence compatibility effect, where participants were faster to press the button following sentences describing compatible handshapes.[7]

But it turns out that people activate motor simulations of specific handshapes even when they're understanding language that doesn't explicitly describe any manual interaction. For instance, think about a sentence like *The young scientist looked at the glass.* To just look at something, you don't necessarily need to move your hand. So the question is whether understanding a sentence like this leads you to engage your motor system to simulate grasping a glass. To find an answer, researchers at the University of Victoria invented a monstrous and unwieldy contraption that looks like something out of the early Paleolithic—the aptly named Graspasaurus, which you can see in Figure 18. The researchers had participants listen to sentences about people looking at objects, like glasses and staplers. They then cued them to grab one of the attachments on top of the Graspasaurus and measured how long this took. Once again, people responded faster when response handshapes were compatible with the actions that one would perform to interact with (e.g., grasp) the mentioned object,

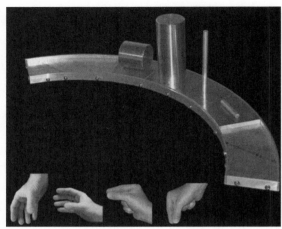

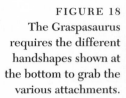

FIGURE 18
The Graspasaurus
requires the different
handshapes shown at
the bottom to grab the
various attachments.

even though, in this case, that interaction was not actually described in the sentence.[8] We learn from this that part of the way that people understand language about objects is by engaging their motor systems to activate the motor routines that might typically be used with them.

The knowledge you have about how to interact with an object is known as that object's *affordances*, and this research with the Graspasaurus shows that you seem to activate affordances representing what you can do with a glass when you see the noun *glass*, and likewise for *belt*, or *grape*, *dish*, and so on. Of course, the specific affordances for each object are quite different; they might involve different actions. But these nouns all describe objects that you can easily imagine interacting with.

So you might be wondering, what about nouns that describe things that are harder to link to bodily action, like *cloud*, *flood*, or *pond*? It's not that you can't imagine at all what you would do to interact with these things; it's just harder than with *glass* or *belt*. If simulating the motor affordances of objects is part of accessing the meanings of words, then could it be the case that it's harder to understand words about things that are hard to imagine interacting with than things that are easy to imagine interacting with?

One way to tell would be by measuring how long it takes people to understand the different types of words. If accessing the meaning for easy-to-imagine-interacting-with nouns like *grape* is easier because the pertinent motor simulation is easier to access, then people should be measurably faster to understand them than hard-to-imagine-interacting-with nouns, like *pond*. To test this, a group of Canadian researchers first had people rate hundreds of nouns for how easy the things they denoted were to imagine interacting with, and they produced two lists of nouns, one that was easy to imagine interacting with and the other that was hard. They were interested in measuring how long it would take people to understand these words. But there are lots of things that can affect how long it takes people to understand a word. So the researchers made sure that the words in their two lists were matched for a variety of these factors, like their frequency, length, and a host of others. As a result, the two lists differed only in how easy it was to envision interacting with the things they denoted. They then ran the actual experiment—they had another group of people simply read the words and had them press one of two buttons as quickly as possible to say whether each word was or was not a real English word.[9] Of course, they had to include some made-up words like *brane* and *ludge* as well, so the answer would sometimes be "no." But they were only interested in their two lists of real words. And when they looked at the data, as you might expect, they found that people's reactions were faster for the easy-to-imagine-interacting-with nouns than the hard-to-imagine-interacting-with ones. The upshot is that when you read nouns, even merely to decide whether they're words on not, you evoke knowledge about how you physically interact with the things that the nouns denote. The easier it is to activate that motor simulation, the faster you're able to understand the word.

At around the same time when the Graspasaurus began to roam the earth, other devices also started popping up that allowed researchers to measure diverse types of manual action. For instance, there's the Knob. The Knob is appropriately named, because it's a knob. It's about one inch in diameter, and it's embedded in a box. It can be rotated sixty degrees in either direction, either clockwise or counterclockwise. In addition, it's wired up to send a signal to a computer when rotated to that limit. Perhaps by now you can guess what kind of action-sentence compatibility effect you could measure using the Knob. Researchers at

the University of Rotterdam had people listen to sentences that implied clockwise rotation, like *Jane started the car*, or counter-clockwise
rotation, like *Bob opened the gas tank*, and then made them rotate the
knob in one direction or the other. Sure enough, people's responses
were faster when the direction of the rotation implied by the sentence
was the same as the one they had to move their hand in.[10]

The Rest of the Body

Thus far, we've only been looking at hand actions, but we know that
there are specialized portions of the brain dedicated to controlling the
different body parts, from the tongue to the toes. Are these parts of
the motor system, which we know come online when people are imagining moving their different body parts, also engaged during language
understanding? To what extent is understanding language a whole-
body process?

We know that the motor system is organized such that sets of neurons that drive very similar actions—like those responsible for executing different handshapes—*mutually inhibit* each other. That is,
when one is active, it sends a signal to suppress activation of the other,
so that your muscles don't get competing, incompatible messages at
the same time. It's critical for actions that use the same body parts to
be mutually inhibitory, because you wouldn't want to try to do two
things with the same body part at the same time, like chewing and
gargling. But if the actions use different body parts, they need not
compete with each other. Although holding a tray and putting in an
earring ought not to be attempted simultaneously, there's no reason to
keep yourself from trying to walk and chew gum at the same time.

We can take advantage of this mutual inhibition of actions that use
the same body parts as a way to tap into how people employ their
motor systems to understand language. Our brain won't allow us to
perform two different actions at the same time using the same body
part, so if language evokes body part–specific motor simulation, then
it should be hard for you to think about an action while simultane-

ously understanding language about another action that uses the same body part. Here's how we tested this idea in my lab.

First, we needed a way to get specific action circuitry fired up. We know from the mirror neuron research (that we saw earlier) that seeing someone perform an action activates the motor system. So we showed images of people performing actions to participants in order to activate their motor representations of those actions. Once these motor representations were active, we presented words describing different actions that use either the same body part or a different one. And we predicted that it would take people longer to process the action words when those words described actions that used the same body part as the action depicted in the image than when the actions used a different body part. An example will probably help here. Suppose you see an image like Figure 19,

FIGURE 19

and then you are asked to decide if the following word describes it:

juggle

It should take you a relatively long time to decide that the picture does not depict juggling, because the two actions use the same body parts (the arm and hand) in different ways. By contrast, you should take less time when you see the same picture and then are asked to decide whether the following word is a good description of the picture:

punt

This is because the word *punt* describes an action using the foot and leg and therefore should not be in competition with the hand action the pictures depicts.

In a series of studies in English and Chinese, we found that it takes people longer to decide that a verb is not a good description of an image when the action it describes and the action the picture depicts use the same body part than when they use different body parts. We infer from this that accessing the meaning of action words involves activating the specific parts of the motor system dedicated to controlling the relevant body parts.

Another way you could test this would be to have people read sentences about actions that use different body parts (*turning a key* is a hand action and *kicking a chair* is a foot action) and then have them push a button with either the same body part or a different one—for these sentences, either their hand or foot. You would expect to see that if the actions that the sentences describe are different from the button-pressing action but use the same body part, then people should take longer to press the button. That is, if someone hears a sentence about *turning a key* and then has to push a button with their index finger, then these two actions should interfere with each other because they use the same body part in different ways. However, if they have to press a button with their foot after hearing the sentence about *turning a key*, there should be no interference, because we can turn a key and flex our foot at the same time, just like we can ride a stationary bike at the same time as we dial a phone. Fortunately, this study has been done. The results: hand sentences slowed responses using the hand, while foot sentences slowed responses using the foot (as you can see in Figure 20).[11]

Let's suppose for a second that we take the preceding evidence at face value, that when people hear or read sentences about foot actions, they engage the parts of their motor system that control movement of the feet and likewise that language about mouth actions engages the mouth-controlling parts of the motor system, and so on. If this is the case, then the topology of the motor system should make a difference to processing. The motor strip is organized by body part,

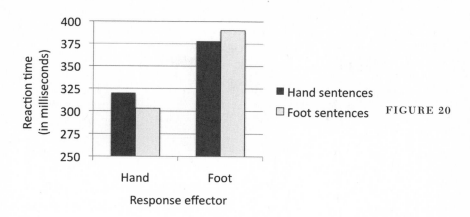

FIGURE 20

such that the mouth-controlling areas are lower down, and as you go up you find those regions that control the hands, and then the areas controlling the feet up at the top. We can use the fact that these different parts of the motor strip are in different places to make predictions about how long it should take people to process words describing actions using the different body parts. This is a little tricky, so it requires some more explanation, but it's also very cool.

As just mentioned, in the motor strip, the neurons that control leg and foot actions are located at the top, with those responsible for hand and arm actions below them, and mouth-controlling neurons are near the bottom. But understanding a word about action naturally involves activating a number of parts of the brain, not just parts of the motor strip. Other neural structures involved include those in areas, for instance, that represent the pronunciation of the word, its spelling, and so on. These other parts of the brain are located closer to the bottom of the motor strip—representing mouth actions—than the top. You can take a look at one proposal for where the networks of neurons responsible for understanding leg-related, arm-related, and face-related action verbs should be in the brain in Figure 21.[12]

What you see from these images is that the clusters of neurons used to understand leg-related words are substantially more spread out than those for arm- or face-related words. This turns out to be pretty important, because when neurons become active and send sig-

leg-related word   arm-related word   face-related word

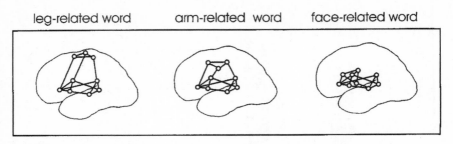

FIGURE 21

nals to other neurons, the propagation of these signals takes time. The farther a neuron sends its signal, the longer it takes for the signal to get there (and for an eventual response to come back). What does all this mean? Well, if understanding words about actions actually engages the parts of the motor strip that control those specific actions, then all other things being equal, people should take longer to recognize and understand leg-related verbs than arm-related verbs, which should in turn take longer than face-related verbs, just because of the distances the neural signals have to travel to activate the entire networks that represent each of these types of word. Crazy though this prediction might seem, it turns out to be true. In several studies, German researchers have found that even when everything else is held equal—for instance the frequency and length of the verbs—leg verbs are processed more slowly than hand verbs, which are processed more slowly than face verbs, as Figure 22 shows.

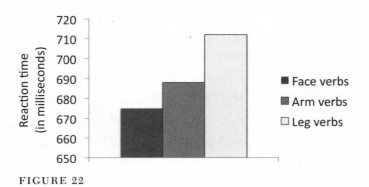

FIGURE 22

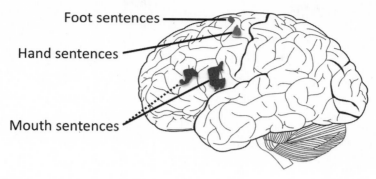

Foot sentences

Hand sentences

Mouth sentences

FIGURE 23

The final piece of evidence that the whole body is involved in understanding language about action comes from brain imaging. Several groups of researchers have looked at whether understanding language about actions measurably activates motor regions of the brain. One group had people simply read hand verbs like *grasp*, foot verbs like *kick*, or mouth verbs like *bite*. Another group gave people sentences like *I grasp a knife*, *I bite an apple*, or *I kick a ball*. Both used fMRI to determine whether the foot, mouth, and hand regions of the motor strip were selectively active during comprehension of the words[13] or sentences.[14] Sure enough, the results came out as shown in Figure 23.

The dark blobs in this image represent those brain areas that were selectively active while people were processing mouth, hand, and foot sentences. There are interesting differences in activation throughout the brain, but of particular interest to us is the activity along the motor strip. As you can see, the lower areas in the motor strip that control mouth actions were selectively active when people were understanding language about mouth actions, while areas a little higher up that control hand actions were active when people were listening to sentences about hand actions, and the part of the motor strip that was most active during foot action sentences was right up at the top, where foot actions are controlled.

Let's come back to the issue that we started with at the beginning of the chapter. Humans, uniquely in the known universe, can take a few characters on a page or a sequence of sound waves and convert them into a mental representation of actions as complex as arm-wrestling techniques. We've now seen some of the mounting evidence that, during this process, people automatically mentally simulate lots of motor details of the described actions, including the direction that they would move in, how they would rotate their hands, what body part or parts they would use, and how their hands would be shaped, and that doing so engages the same brain systems as performing real actions. The motor systems of the brain appear to be working overtime.

What's more, to come back to an issue we discussed in the last chapter, the fact that people perform motor simulation while understanding language is yet another piece of evidence that people are creating scenes in their minds from the perspective of an immersed experiencer. Motor simulation is intrinsically about projecting oneself into a body—often someone else's—and, when you simulate what it would be like to do things someone is described as doing, you're taking their perspective, not merely in a visual way, but in terms of what it would be like to control their actions. Understanding language, in multimodal ways, is a lot like being there.

CHAPTER 5

More Than Words

There's a dilemma in cognitive neuroscience known as the *binding problem.*[1] Suppose you're playing Pac-Man. You're doing your best to eat the dots and the bonus fruit while avoiding Pinky and Blinky and the rest of the bad guys. When you look at the screen, you can see your Pac-Man avatar moving—cruising around, moving upward, then to the left and off the screen, only to reappear on the far right, and so on. And at the very same time, you see its color and characteristic Pac-Man shape. Shape, color, location, and motion all the same time. And likewise, you see the bad guys, Inky and Clyde and company, going their own ways, with their own shapes and colors. And each of these percepts—whether of Pac-Man or Inky—is coherent. And by "coherent" I mean that even though you see both Pac-Man and Inky at the same time, you know that it's the yellow Pac-Man that's moving upward while the blue Inky is moving downward. The binding problem is the question of how your brain manages to mechanically put that information together in the right way. We know that the way your visual system works is to separate out the different properties of the objects you see into discrete streams. The location and movement of the objects are processed in the Where pathway, running from the occipital lobe up through the parietal cortex, while the objects' shape and color is processed in the What pathway. How are these distinct processes in different collections of neurons hooked back up together in the right way? In other words, if you activate distinct representations of the color, shape, location, and movement of several objects at the same time, how are you able to bind together Pac-Man's yellowness, shape, location, and movement, while at the same time binding none of these with Inky's blueness, ghost-like shape, location, or direction (which all need to be bound together themselves)?

I bring up the binding problem not to offer a solution to it here but because it serves as a useful analogy for something that comes up in language, which has its own version of the binding problem.[2] Suppose instead of playing Pac-Man yourself, you're listening to someone narrate their ongoing Pac-Man game, and you hear them say *That blue ghost is about to catch my Pac-Man!* Each of the words uttered makes its own contribution to what you will eventually mentally simulate: *blue*, *ghost*, *catch*, and *Pac-Man* evoke colors, shapes, and actions that you can load up into your simulation. But how do you know how to fit them together? How do you know that the ghost is blue and not Pac-Man. (Admittedly, if you know a lot about Pac-Man to begin with, then you already know that Pac-Man is yellow, but you can still track which object is what color and moving where even if you're a Pac-Man novice.) How do you know that Pac-Man is about to be caught by the ghost, and not the reverse? So the linguistic binding problem is this: how do you assemble the contributions that the various words make to simulation so that the right parts go together in the right way? Asked another way, if words provide the cast of characters, props, and sets, what directs the action so that the scene plays out in the intended fashion?

Assembling words to drive simulation is a complicated operation. But although the binding problem is still hotly debated in the literature on cognitive neuroscience of perception, we actually have a pretty good grasp on how language solves its version of this problem. Words are obviously a critically important factor in simulation—they act as pointers to the detailed experiences to be evoked in simulation.

But there's more at play in utterances than just words. To know who's yellow and who's blue and who's chasing after whom, to know where to position your mental camera and what to focus on, you need something more than words. You need grammar.

Sentences as Plants

Linguists have long known that most of a sentence's meaning resides in words and that those words have to get assembled into longer bits. The analogy that linguists like to use is to liken a sentence to a plant. Words are leaves—the meaty, fleshy, part—and grammar is the system of branches and stems that holds them in place. Grammar puts sentences together. It's in charge of their form, like the order that words appear in (it ensures that the subject *That blue ghost* comes before the verb *is*, for instance), and even things like subject-verb agreement (*The blue ghost is* rather than *are*). These aspects of the form of a sentence are important. They distinguish a string of words that form a viable utterance in a language from a string that's mere gibberish. Randomize a bunch of words, and you get something that's unrecognizable as an English sentence, like *blue that is Pac-Man my about catch ghost!* Not only is this unrecognizable as English, it's also uninterpretable. The form of a sentence is a large part of what makes its meaning accessible.

How exactly does grammar help make meaning? Let's again take the sentence *That blue ghost is chasing my Pac-Man!* When you read it, you know that *blue* is a property of the *ghost* and not *Pac-Man*. You know that because, as a solution to the linguistic binding problem, English has a grammatical rule whereby when you put an adjective (like *blue*) right ahead of a noun (like *ghost*), the property that the adjective describes applies to the entity that the noun denotes. This is actually an example of a much more general principle that grammar works with. It's easier to keep track of what needs to be bound to what if you organize words together into larger groupings. All three words in *That blue ghost* apply to the same entity, in the same way that the words in *my Pac-Man* both apply to a single entity. Linguists have a name for a grouping of words like this that's structured around a noun with other bits that modify it, like adjectives, possibly adverbs, and maybe articles like *the* or pronouns like *my*. This is a noun phrase. (It's a *phrase* because it can have a number of words, and *noun* because that's the most important part.) These combinations of words form units of meaning; the semantic contributions of the words in each noun phrase remain—for the most part—local to that noun phrase. That's how you know that *blue* applies to *ghost* and not *Pac-Man*.

But that turns out to be only the beginning of what grammar does to bind the parts of a sentence together. Expanding our view from just the narrow scope of noun phrases to the whole sentence, we might ask: How do we fit together the meaning of the verb with these noun phrases hanging out around it? In this example, you know stuff about the verb *chase*. For instance, it describes an event that typically involves two entities playing distinct roles—let's call them the "chaser" and "chasee." Conveniently, the sentence mentions two entities—*That blue ghost* and *my Pac-Man*. So far so good. But is *That blue ghost* supposed to be the chaser or the chasee? And how about *my Pac-Man*? To talk about a chasing event, it's obviously not sufficient to use the word *chase* and mention the entities that fill its two roles. You also need some means to convey which entity plays which of the verb's roles. This is another thing that grammar does—it indicates which parts of a sentence fill which roles. The way it does this is by mapping specific parts of the form of the sentence, like the order that you mention entities in, onto the verb's roles. In simple *transitive* sentences in English, like the one we're looking at, the entity mentioned first will be the one performing the action—in this case, the chaser—while the entity mentioned after the verb will be the one the action is performed on—the chasee. As a speaker of English, this is a rule you know. This kind of knowledge about how the global structure of a sentence maps onto roles that mentioned entities play in an event is known in the linguistics literature as an *argument structure construction*, and the particular one in play in this sentence is a transitive construction.

That blue ghost	*is about to catch*	*my Pac-Man*
The rabid monkey	*is devouring*	*the unconscious scientist*
Noun Phrase 1	**Verb**	**Noun Phrase 2**

FIGURE 24 The transitive construction, with examples.

Now, to my knowledge, no studies have explicitly investigated how grammatical configurations, like being the subject of a verb or occurring within a noun phrase with another word, affect embodied simulation. This is most likely because it goes almost without saying that if people perform embodied simulations in understanding sentences, they perform different simulations for *That blue ghost is about to catch my Pac-Man* and *My Pac-Man is about to catch that blue ghost.* Nevertheless, there is some indirect evidence that the way grammar aligns the semantic contributions of individual words is reflected in simulation. For instance, recall the action-sentence compatibility effect we saw in the last chapter, in which people move their hands faster when they're performing a movement compatible with the action described in the sentence they're responding to. Well, when we reconsider the stimuli used in these experiments from the perspective of grammar, we see that what the subject and object are can make all the difference to simulation. For instance, the difference between *Alex forked over the cash to you* and *You forked over the cash to Alex* is in whether *you* or *Alex* is the subject, and this is expressed by where in the sentence each of these is placed. And yet this difference is sufficient to drive embodied simulation of actions in opposite directions—*Alex forked over the cash to you* speeds up hand movements toward you, but *You forked over the cash to Alex* speeds hand movements away from you. The clear implication is that the motor action understanders simulate depends not only on the action described, but on what roles the mentioned event participants play in that action—as indicated by grammatical cues like subject and object.

Grammar contributes to simulation by putting together all the pieces contributed by words in the right configuration. You use the order of words and other grammatical markers to figure out what properties should go with what entities and who should be simulated doing what to whom. But combining the simulatable content from individual words is only one of grammar's contributions.

Sentences as Plants, Revisited

Let's return to the analogy I mentioned earlier about how words and grammar work. Linguists view sentences as like plants, in that words are leaves hanging from the stems of grammar. You can actually get a lot of mileage out of this analogy. Just as leaves perform the primary life function of photosynthesis, so words perform the primary linguistic function of conveying meaning. And just as stems and branches play the support role of holding leaves in place and linking them to the rest of the plant, so grammar supports meaning by placing words in linear order and linking them to the rest of the utterance. In the preceding section, we saw that grammatical structures are responsible for combining words into meaningful utterances, thereby constraining and configuring the simulation contributions that individual words make.

This separation of labor, between functional units (words or leaves) and their supporting lattices (branches or grammar) is extremely appealing. But, when we push the botanical side a bit, it turns out to be an oversimplification, at least for plants. It is true that in most plants, leaves are responsible for many critical life functions, most notably photosynthesis. And yet, as it turns out, stems are often not content to serve as mere handmaidens to their photosynthesizing masters. Stems, even the woody ones typical of trees, perform some measure of photosynthesis.[3] In some plants, like cacti, stems actually perform the bulk of the photosynthesis. And this is where the analogy between plants and sentences gets really deep. The same is true of grammar. Like branches, grammatical structures serve not only to arrange the forms and meanings of words; they can also contribute meaning of their own.[4] In some cases, their meaning contributions are vague or abstract. In other cases, the broad-stemmed cacti of the grammar world, grammatical structures can be responsible for a large part of an utterance's meaning.

We'll begin with argument structure constructions,[5] patterns that arrange and bind together verbs and the other parts of a sentence. There are lots of them, beyond the humble transitive. A frequently discussed example in English is the *ditransitive*, which you see at work in sentences like *John sent his landlord the check* or *The defender kicked his goalie the ball*. This structure has a particular form, as seen below. It starts with a noun phrase, followed by a verb, and then two more noun phrases. And here's the thing. In addition to this form, the ditransitive comes with a particular meaning. Namely, sentences that use the ditransitive, like *The defender kicked his goalie the ball*, describe scenes in which the first entity (*the defender*) transfers the third entity (*the ball*) to the second entity (*the goalie*). Said more concisely, ditransitive sentences have a *transfer-of-possession* meaning.

The defender	*kicked*	*his goalie*	*the ball*
John	*sent*	*his landlord*	*the check*
Noun Phrase 1	Verb	Noun Phrase 2	Noun Phrase 3

FIGURE 25 Examples of ditransitive sentences.

Although it's clear that ditransitive sentences like the ones in Figure 25 do indeed describe transfers of possession, it's reasonable to be skeptical about whether this meaning actually comes from the argument structure construction. After all, it could just as well be the verb, right? The verb *send*, for example, seems like it's mostly about transferring objects. So maybe the meaning of the sentences as a whole is just a combination of the parts that make it up, and the transfer-of-possession meaning comes from the verb. No need to rely on an argument construction to contribute that. The thing is that when we look a little at the range of verbs that appear using the ditransitive pattern, we find that they include a whole host of other verbs that don't describe only transfer events. For instance, *kick* doesn't necessarily denote transfer—consider typical uses like *Don't kick your brother* or

Stop kicking—you'll fall out of your chair! Only when you use *kick* with ditransitive grammar context, as in *The defender kicked his goalie the ball* does it gain a transfer-of-possession meaning. The use of verbs like *kick* in sentences like this indicates that it's possible for verbs that don't necessarily denote transfer to do so when given the right, ditransitive argument structure pattern.

Nevertheless, we still might not be convinced that the ditransitive grammar is the thing responsible for the transfer-of-possession meaning. It could in principle be that expert users of English like you have actually learned multiple meanings of *kick*. One of them denotes an action on some object with the leg or foot, as in *Don't kick your brother*. And the other denotes such an action performed in such a way that the kicked object is transferred to some third party, as in *The defender kicked his goalie the ball*. It could be that when you use *kick* ditransitively, you're merely picking the second of these meanings, and again, the grammar doesn't contribute any meaning at all.

How can we tease apart whether it's the verb or the grammar that provides a particular meaning? One way is to turn to verbs for which you cannot possibly have memorized a transfer-of-possession meaning. Consider, for example, the verb *to motorcycle*, as in *The delivery boy motorcycled his client some blueprints*. To my knowledge, and hopefully to yours as well, this is a totally novel use of this verb. If you were ever to use *motorcycle* as a verb, which is probably unlikely to begin with, you'd probably use it meaning "to change locations using a motorcycle," as in *We motorcycled down the coast*. But I'd be willing to bet that you've never heard anyone talk about "motorcycling someone something"! Yet, you have little trouble interpreting *The delivery boy motorcycled his client some blueprints*—it clearly describes a transfer of possession from the delivery boy to the client of some blueprints. Surely, in cases like this one, the transfer-of-possession meaning must be contributed by the ditransitive construction itself and not the verb. And when you look around, you'll find bountiful examples like this. The typical English speaker has never used *tennis racket* as a verb. And yet, when confronted with an utterance like

Venus tennis racketed her sister the hair clip, you know that Venus transferred the hair clip to her sister by means of the tennis racket. In cases like this, the verb again can't be responsible for the transfer-of-possession meaning because it's not a verb to begin with; instead, it has to be contributed by the ditransitive grammar itself.

If you were still skeptical of the claim that it is an argument structure construction—the ditransitive—that provides the transfer-of-possession meaning to these utterances, you could maybe retreat to a weaker position. Maybe you're able to interpret these sentences because of a more general capacity for making inferences. That is, maybe when you put noun phrases describing three entities (one that happens to be capable of causing a transfer, one capable of being transferred, and one capable of receiving a transferred object) in a sentence, the most natural thing for someone to do is to try to make the best sense out of the sentence that they can. And so under these unusual conditions, people use their knowledge about the world—in particular about the particular entities mentioned by the noun phrases—to figure out what type of event could have involved them all.

But there are several reasons to believe that general inference processes aren't enough, and that people really do bring specific grammatical knowledge to bear on interpreting these sentences. For one, switching the order of the noun phrases following the verb results in sentences that are either meaningless or describe a transfer in the reverse direction. *The delivery boy motorcycled some blueprints his client* is hard to figure out unless one assumes anthropomorphic blueprints and a transferable client. Second, native speakers of many other languages deem sentences of the same type to be ungrammatical. For instance, the French equivalent of the sentence *John sent his landlord the check* is completely unacceptable: no Frenchman in his right mind would accept *Jean-Luc a envoyé son propriétaire le cheque* as being well-formed French. (The best possible interpretation of this might be that John sent his landlord, whose name is The Check to somewhere unspecified.) In sum, because the order matters in ditransitive sentences and because their use seems to be specific to particular languages, like English (and Thai and Chinese), but not others like French (and German and Japanese), it's quite clear that language users are not simply relying on general inference abilities but rather are using knowledge of a specific grammatical pattern that encodes the transfer-of-possession meaning.

There is also experimental evidence. One study from the University of Wisconsin used several different tasks.[6] In the first experiment, people performed a sentence choice task, in which they saw pairs of sentences, such as the ones below. Each sentence in the pair has roughly the same set of words, but they use different argument structure constructions. The first sentence uses the ditransitive we've been working with (*Lyn crutched Tom her apple*) and an additional phrase (*so he wouldn't starve*). But the second uses a different construction, the transitive construction that we saw earlier (*Lyn crutched her apple*) also along with an additional phrase (*so Tom wouldn't starve*). Importantly, in these sentences, the verb (here, *crutch*) is not much of a verb at all. *Crutch* is primarily a noun, but, because English allows some nouns to be "verbed," it can be forced into service as a verb.

Lyn crutched Tom her apple so he wouldn't starve. [ditransitive]
Lyn crutched her apple so Tom wouldn't starve. [transitive]

The researchers had people read both sentences and then look at one of two inference statements, like the ones you see below. One of the statements was consistent with the purported transfer-of-possession meaning of the ditransitive, while the other was consistent with the purported meaning of the transitive grammar, which as we saw is used to describe one entity acting on another. And they asked the participants to indicate which of the two sentences above most strongly implied that the inference statement was true.

Inference statements:
Tom got the apple. [transfer-of-possession]
Lyn acted on the apple. [act-on]

If language users take into account the meanings of argument structure constructions when interpreting a sentence, then they should choose the member of the sentence pair consistent with the inference statement. For instance, they should decide that the ditransitive *Lyn crutched Tom her apple so he wouldn't starve* more strongly implies that *Tom got the apple.* And because the verbs are derived from nouns and do not therefore have transfer-of-possession meanings to begin with, any such meaning found in the sentence must arise from the grammar, not the verb.

As predicted, participants displayed a clear and very strong preference. Ditransitive sentences more strongly implied the transfer-of-possession meaning. Participants picked the ditransitive member of a pair 80 percent of the time. This strong preference indicates that the presence of the ditransitive construction in a sentence leads understanders to think that a sentence is about transfer of possession.

But a single experiment rarely settles a research question definitively, and this one is no exception. As with most experiments, the design of this experiment left open a possible alternate explanation for the results that needed to be teased apart in subsequent work. Namely, it's possible that it was something about the experimental task itself that caused people to prefer one sentence or the other for each inference statement. The experiment explicitly presented participants with a description of the hypothesized transfer-of-possession meaning of the sentence, and asked them to make judgments about sentences. But it could be that people don't typically think of ditransitive sentences as indicating transfer—it only occurs to them when they're prompted to, as in a task like this. In a second experiment, the same researchers tested whether the preference they observed in the first experiment would appear without explicit prompts. So they presented the same sentences, and asked people to just paraphrase either the critical sentence or the verb. It turned out that participants were significantly more likely to provide a transfer-of-possession paraphrase for ditransitive sentences than a transitive one. For instance, they might paraphrase the ditransitive sentence as "Lynn poked at the apple with the crutch to get it to Tom" but the transitive sentences as "Lynn poked at the apple with the crutch to break it into pieces." They were also more likely to define the verb as denoting transfer of possession when it was used in a ditransitive sentence—something like "to use a crutch to move something to someone." We learn two things from these studies. First, people use grammatical constructions like the ditransitive to interpret the meanings of sentences as a whole. And they use them for a second purpose as well. They also use the meaning of an argument structure construction to make inferences about the meanings of the verbs used with them.

The idea that grammatical constructions might be meaningful opens a Pandora's box. How many grammatical constructions are there? How many different types of sentence have their own specific meanings? Some linguists have gone so far as to argue that every grammatical construction contributes some meaning or function to the utterance it appears in. The idea is that for every difference in linguistic form between two utterances—whether the order of the words or the markers on them—there is a concomitant difference in their meaning or function. This is known as the *principle of no synonymy*, or the *principle of non-equivalence of grammatical forms*.[7] Like most principles, it's much easier to prove this tenet wrong than right; the former requires evidence of only one grammatical difference that does not affect meaning or function, while the latter requires an exhaustive demonstration that all such formal differences result in meaning differences. Nonetheless, case studies are progressively accumulating, which show at least that many grammatical differences yield meaning differences.

Here's an example of a really subtle case. There happens to be another grammatical construction that closely approximates the ditransitive in both form and meaning—the *caused-motion construction*. The caused-motion construction, seen in sentences like *John sent his check to the landlord* or *The defender kicked the ball to his goalie*, differs from the ditransitive in terms of its form in just two ways. First, instead of expressing the goal of the action with just a noun phrase (like *the landlord* or *his goalie*), it inserts a preposition at the beginning to make *to the landlord* or *to his goalie*. Second, the order of the two entities mentioned after the verb is reversed—rather than recipient then object as in the ditransitive, the caused-motion construction mentions the object then the recipient. Some theories of syntax claim that there are so many similarities between ditransitive and caused-motion sentences that they ought to be treated as superficially different versions of the same underlying structure.[8] And although this account is appealing for its simplicity and parsimony, the two constructions differ in enough ways that they can be shown to clearly be distinct. For one, there's a range of differences between them in how they're used, including how formal the discourse is, which of the entities after the verb is new to the discourse, and which is longer.[9] Another key difference between the ditransitive and the caused-motion constructions is in the verbs that they allow to appear in them. It appears that the caused-motion construction is slightly more permissive than the ditransitive is in this regard, such that some verbs are perfectly acceptable only with the former—according to many native English speakers, it's fine to *contribute some time to a cause*, but unacceptable to *contribute a cause some time*.

But do the two constructions also differ in meaning? Do the embodied simulations driven by these two grammatical constructions differ at all? There are subtle differences in how the two constructions are used, ones you've probably never noticed before. For instance, it appears perfectly acceptable to suggest that you will *throw your keys to the floor* but it is strange to say that you will *throw the floor your keys*. Most native speakers of English have the intuition that the first, caused-motion version can indicate that you moved the keys to a new location—the floor—by means of throwing them. By contrast, in order for the ditransitive version, *throw the floor the keys*, to make sense, the word *floor* has to refer to some set of relevant people associated with the floor—perhaps the people currently standing there. And you get the same asymmetry between the ditransitive and caused motion constructions when you compare *The outfielder hit the fastball to the left field wall*, which should sound fine, with *The outfielder hit the left field wall the fastball*, which should only sound reasonable if *the left field wall* stands for someone located there, like maybe fans leaning over it.

So it appears to be that despite superficial similarities, the ditransitive and the caused motion constructions actually differ in terms of the meanings they convey. The ditransitive describes an intended transfer of the object to a recipient. And naturally, a recipient must be capable of receiving something. A floor or a wall isn't a good recipient, and so to interpret a ditransitive sentence in which a floor or wall is placed in this role, we have to use some other means to interpret *floor* or *wall* such that they do indicate a recipient. By contrast, the caused-motion construction appears to describe scenes in which some entity causes another entity to move along a path. A floor or wall can be end-points of such a motion, as can any location, so the caused-motion versions of these sentences are perfectly easy to interpret.

There's additional linguistic evidence for this distinction between the caused-motion and ditransitive constructions. If a caused-motion construction really conveys a meaning of caused motion along a path, then it should be compatible not only with the preposition *to*, but with any other similar preposition. So we can not only *bring the ladder to the wall* but also *lean the ladder against the wall*. We can not only *carry coals to Newcastle* but also *place the remains in the ground*. However, in agreement with its claimed transfer-of-possession meaning, the ditransitive doesn't allow these configurations. In fact, it is quite difficult to imagine what it would mean to *lean the wall the ladder* or *place the ground the remains*!

Let's turn to the question of whether embodied simulation differs when you hear them. What predictions might we make? If the caused-motion construction highlights the motion of an object along a path (while the ditransitive does not, focusing instead on the transfer of possession), then when you hear caused-motion constructions, you should construct more detailed embodied simulations of the described path of motion. For instance, one difference between the caused-motion *You are sliding the cafeteria tray to Sally* and the di-

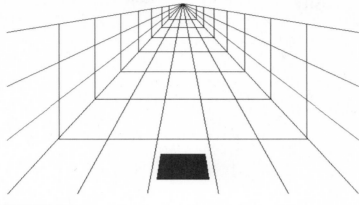

FIGURE 26

transitive *You are sliding Sally the cafeteria tray* might be that the caused-motion version causes you to perform more detailed embodied simulation of the motion of the tray away from your body, and toward Sally, than the ditransitive version does.

We actually tested this hypothesis in our lab. The logic went like this. Suppose we have people listen to caused-motion or ditransitive sentences that describe moving an object along a path with your hand. And then right afterward, we have them move their own hands along a different path. If the caused-motion sentences lead to more simulation of their paths, then they should interfere more than the ditransitive sentences with the subsequent actual hand motion.

This is how we set it up. First, we had people look at a grid on a computer monitor, like the one in Figure 26. They had to click on the dark box in the foreground to hear a sentence. Then the sentence played aloud. It used either the caused-motion or ditransitive construction, describing *you* moving an object away from your body, just like the Sally sentences. Right after each sentence, a bull's-eye appeared somewhere on the grid, and the participants had to move their hand to click on it. If caused-motion sentences drive people to construct more intense or detailed simulations of the path of motion, then this should affect the hand movements they perform after hearing caused-motion sentences about hand motion more than with ditransitive

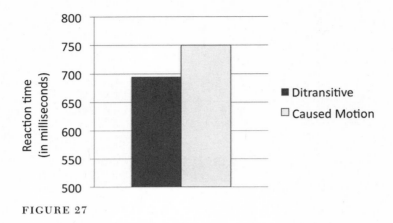

sentences. In particular, if they're mentally simulating moving their hand along a path to understand a caused-motion sentence, then when called upon to subsequently move their hand along a path, they should do so for a longer time.

As you can see from the results in Figure 27, this is what we found—people moved their hand for longer after caused-motion sentences than ditransitive ones. So caused-motion sentences appear to produce more detailed embodied simulation of the path of motion than ditransitives. Different argument structure constructions really do seem to contribute meaning to be mentally simulated.

It appears that the analogy between plants and utterances is even more apt than previously thought. Just as stems can photosynthesize, so grammatical constructions can convey meaning—meaning that is directly fleshed out in concrete components of an embodied simulation.

How Grammar Modulates Our Perspective

There's actually a third way that grammar contributes to simulation. It also tells you not <u>what</u> to simulate, but <u>how</u> to simulate it. Even if you know what entities and events a sentence describes, some properties of embodied simulation are nevertheless underspecified. For instance, take a sentence like *The driver gripped the steering wheel and pounded on the horn as he waited for the light to change.* What perspective do you adopt when you mentally simulate this event? The perspective of the driver? Or of an outside observer? You can tell which you chose because the different perspectives lead to very different simulations. If you're simulating yourself as the driver, you'll be projecting your own visual system into that of the driver, so that you see what the driver would see; a steering wheel, with hands gripping it, in front of a windshield through which the street and perhaps a stoplight are visible. Creating a simulation from the perspective of someone taking part in an event takes a *participant perspective.* But you can also simulate an event from an *observer perspective.*[10] If you're not the driver but merely an observer, you can place your imagined eye—the mind's camera—at any of a number of other locations. For example, it could be placed directly behind the driver—this would be nearly identical to the participant perspective, except that it would also include the back of the driver's head. Or it could be placed outside of the car to one side or in front, so that it includes the car, along with the driver. Or it could be a view from the front of the car that sees the driver through the windshield. Which perspective you adopt is important, because it will radically affect your embodied simulation of an event. And you don't just pick a perspective at random.

Research on which perspectives people adopt and why has mostly focused not on language but on memory. As it turns out, you're more likely to adopt a participant perspective when recalling events that you have a positive attitude toward (like your favorite birthday memory), rather than ones you have a negative attitude toward (like your least favorite birthday memory). You're also more likely to adopt a participant perspective when you judge yourself to have changed little since the recalled event—for instance if you're recalling going to your childhood church and you're still religious—than when you feel you have changed a lot since then.[11] At the biological level, visual imagery from these two perspectives differs both in where the neural structures that support it are located and in which brain hemisphere dominates its performance.[12] Now, when comprehending language, there's more than just your relationship to the described event in play. You also have the utterance itself. And as we'll see, it seems like grammar can push your perspective around, too.

Let's return to the sentence we saw above—*The driver gripped the steering wheel and pounded on the horn as he waited for the light to change*. When people first read a sentence like this, more often than not they adopt an observer perspective; they see the driver gripping, pounding, and waiting from an external vantage point. But a tiny grammatical modification in the sentence can change that. Suppose we modify the subject of the sentence to make it *You gripped the steering wheel and pounded on the horn as you waited for the light to change.* If you are like most people, you probably have the intuition that the second-person subject *you* makes you more likely to adopt a participant perspective. This observation leads to an intriguing idea. Maybe grammatical person (first person *I*, second person *you*, third person *he* or *she*) orchestrates the perspective you simulate from.

We ran a series of experiments in our lab to investigate this question. We wanted to find a set of events that would look different depending on the perspective they were viewed from. And we came up with events in which an object moves away from someone. When you move an object away from you, it appears to get smaller. But when you see someone else move an object away from them, at least when you see the action from the side, it changes position laterally without necessarily changing in apparent size. So the same event, an object moving away from someone, looks different depending on whether you are that someone or not. And we reasoned that the same should be true when you're simulating objects moving away from you versus from someone else. So if second-person language about objects moving away from *you* (say, *You threw the baseball to the catcher*) leads people to simulate from a participant perspective, then they should simulate the objects getting smaller. But if third person language (*The pitcher threw the baseball to the catcher*) leads people to simulate from an observer perspective, then they should be simulating objects being displaced laterally but not necessarily changing in size.

Here's how we tested this idea. We had people listen to sentences describing motion away from someone, like someone throwing a baseball, and gave them either a second-person subject (*you*) or a third-person subject (*the pitcher*). These sentences were interleaved with a second task, which participants thought was totally unrelated. Immediately following each sentence, two pictures popped up on the screen in quick succession, and participants had to press buttons labeled "yes" or "no" to indicate whether the two pictures depicted the same object. Sometimes the two objects were the same (say, two baseballs in a row), and sometimes they weren't (a baseball and a shoe). We were only interested in what happened when the two pictures depicted the same object. And here's the catch. When the two objects were the same, we manipulated how the second of the two images looked, so that in combination with the first image, it looked like the object was either moving away from the participant, or from their left to right. The way we did this was pretty simple. The first image appeared briefly at the center of the screen, followed by an even shorter visual mask (visual static, as you can see in Figure 28). Then the second image appeared, either offset a little to the right and the same size as the first image (top row in the figure), or at the center of the screen and slightly smaller in both height and width than the first image (bottom row). The figure shows a stop-motion version of the rightward-moving condition and the away-moving condition. The trick is that because the images were each presented for just half a second, they produced the optical illusion of a single object moving either to the right or away from the participant.

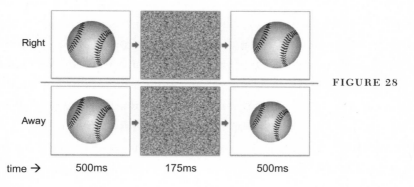

FIGURE 28

Right

Away

time →   500ms   175ms   500ms

Research we looked over in Chapter 3 using a similar method had previously shown that sentences with second-person pronouns lead to faster responses to images when the objects move in the direction "you" would see them moving in.[13] For instance, *You threw the baseball to the catcher* facilitates responses to objects that appear to be moving away. We anticipated that we would be able to replicate this finding in our sentences that had *you* as a subject. But why did we think that people would simulate externally observed motion as going from left to right? An observer perspective simulation of a scene like *The pitcher threw the baseball to the catcher*, for example, can adopt one of a number of different vantage points—from behind either of the participants, from the side, from above, and so on. However, there's actually some research showing that English speakers most commonly simulate horizontal motion as progressing from the left to the right.[14] (Why they do this is a whole different story, but it might have something to do with the direction that we write in, as we'll see in Chapter 8!) So we reasoned that, when native English speakers mentally simulate motion from an observer perspective, that motion should tend to progress from left to right. So in the experiment, we contrasted motion that appeared to be moving away from people with motion to their right. If people adopt a participant perspective when *you* is the subject, they should categorize objects moving away from them faster, and if they adopt a participant perspective when someone else is the subject, categorization of rightward moving objects should be faster.

And that's precisely what the data in Figure 29 show. People responded significantly faster when the perspective induced by the grammatical person and that of subsequently moving images were compatible. This suggests that person acts as a grammatical cue that modulates the character of embodied simulations; in particular, that who is described as engaging in an activity affects the perspective we're more likely to adopt in mentally representing that activity.

Experimental results are always most convincing when they're convergent—when other work, done in other labs, using different meth-

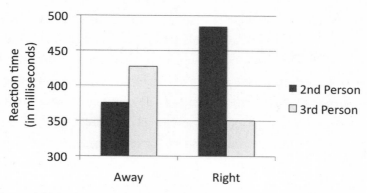

FIGURE 29 A significant interaction of grammatical person with direction of picture motion.

ods, points to the same conclusions. Researchers at Tufts found a related effect using a very different methodology.[15] In a pair of studies, they had people read sentences with different subjects *I*, *you*, or *he*. The sentences described hand actions like *slicing a tomato* or *ironing pants*. Then after each sentence, people saw a picture, taken from either the perspective of the doer (a participant perspective picture) or from across from the doer (an observer perspective picture). Their task was to decide whether the picture showed the action described in the preceding sentence. (In other trials, not of particular interest here, the sentence and the action didn't match in terms of their action; it could be a sentence about slicing a tomato, and a picture of something other than slicing a tomato.)

What they found was that people were faster to say that the participant perspective picture matched the action following sentences about *you*, but they were faster with the observer perspective picture following sentences about *he*. This reinforces the finding from the object motion direction study in Chapter 3.

But there was an interesting, additional result from this work. What perspective would people adopt when they read first-person sentences—sentences about *I*? Would these sentences work more like *you* sentences, evoking a participant perspective, or like *he* sentences, evoking an observer perspective? It might go either way—when someone is talking about himself or herself, you can simulate the things they're describing from either their perspective or an outside perspective. And that's what the researchers found. In a first experiment, people responded faster after *I* sentences to the participant perspective pictures than the observer perspective pictures. That is, they simulated what it would be like for them to be in the role of *I*. But, in a second experiment, the researchers added a little context to the mix. They preceded each sentence with a setup, describing who *I* was, for instance:

> *I am a 30-year-old deli employee.*
> *I'm making a vegetable wrap.*
> *Right now, I am slicing a tomato.*

In this setup, people were faster to respond to the observer perspective pictures after *I* sentences. That is, they were simulating someone else slicing the tomato. One explanation for this is that people are more likely to adopt an observer perspective when they hear *I* and have a concrete image of the person who's speaking or writing. The more detail the context gives about that person, the more able the comprehender is to represent what that person is like. By contrast, they're more likely to think of *I* as referring to themselves when they don't have any idea who the writer or speaker is.

The bigger point here, though, is that grammatical person seems to modulate the perspective you adopt when performing embodied simulation. This isn't to say that every time you hear *you*, you think about yourself as a participant in simulated actions. But it does mean that the grammatical person in a sentence pushes you toward being more likely to adopt one perspective or another. What's interesting about this is that in this case, grammar appears not to be telling you what to simulate, but rather, how to simulate—what perspective to simulate the event from. Instead of acting as the script in this case, grammar is acting as the director.

Grammatical Aspect Modulates Simulation Focus

What's the difference between saying that *The alarm clock was ring-ing* and *The alarm clock had rung*? Or between *The children were lin-ing up for lunch* and *The children had lined up for lunch*? All these sentences describe situations in the past using the past tense (because *was, were,* and *had* are all past tense forms of their respective verbs). But at the relevant time in the past, the events they describe appear to be at different stages. In the scene described by *The alarm clock was ringing,* the alarm clock is in the middle of the act of ringing. By contrast, with *The alarm clock had rung,* at the point in time de-scribed by the sentence, the alarm clock is no longer ringing.

Grammatical distinctions like these, among different framings or views on the same event, fall under the category of *grammatical aspect.* Broadly speaking, whereas *tense* provides an indication of when a de-scribed event takes place (in the past, the present, or future), *aspect* marks the structure of the event—whether it is ongoing, completed, beginning, and so on.[16] Every human language encodes grammatical aspect of one type or another. The most widely discussed aspectual distinction in English (see the examples below) is between the *pro-gressive,* which linguists argue accentuates the internal structure of an event, and the *perfect,* which they claim encapsulates or shuts off ac-cess to the described process, while highlighting the resulting endstate. Native speakers, even those untrained in linguistic analysis, usually agree with these intuitions when you ask them to decide whether events are completed[17] or to match pictures to sentences.[18]

> *John is closing the drawer.* [progressive: highlights internal struc-ture of the event]
> *John has closed the drawer.* [perfect: highlights the resulting end-state of an event]

Behavioral evidence using finer tools points to the same conclusion. For instance, one study had people read narratives several sentences long, in which one critical sentence used either the perfect or the progressive.[19] Then after they were done reading, they performed a memory test, in which they saw a description of that same event but without tense or aspect marking (e.g., *close drawer*) and had to decide whether that particular event had been mentioned in the narrative. The participants turned out to be significantly faster to say that the event had been mentioned when it had originally appeared in the narrative with progressive, rather than perfect aspect. This seems to indicate that the progressive does indeed allow greater access to—or activation of—a representation of the described event. And other work demonstrates that you get the same effect if you have people try to recognize the entities involved in mentioned events—it's easier to remember that *John* was mentioned in a narrative if he showed up in a sentence using progressive rather than perfect aspect.[20]

Complementary evidence on the function of the perfect comes from other work, which found that perfect sentences increase end-state focus, when compared with their progressive counterparts.[21] In that work, participants read either progressive or perfect sentences, then saw an image depicting the event in an ongoing state (e.g., a drawer being closed) or a completed state (e.g., a drawer completely closed). The experimenters found that participants responded to completed-state pictures faster than ongoing-state pictures following perfect sentences. This suggests that the participants were not representing the internal structure of events described with the perfect, so much as their resulting endstates.

These findings compellingly demonstrate that progressive aspect increases access to or activation of the internal components of described events and that perfect aspect does the same for the resulting endstates of events. It seems natural, then, to ask how this difference in access or degree of activation is fleshed out in differences in embodied simulation. It's known that embodied simulations can differ in the degree of detail, or granularity, with which particular parts of scenes are simulated.[22] If I ask you to imagine your kitchen sink, your mental image might not at first include such details as the color of the molding around the basin or the shape of the drain mouth. But when called upon to mentally represent these details, the mind's eye can increase the zoom of the lens through which the imagined object is viewed, and the literature on intentional mental imagery is rich with demonstrations of this.[23] So a simulation-based account of the function of aspect might predict that sentences with progressive aspect should yield greater embodied simulation of the process, or nucleus, of the described event than should corresponding perfect descriptions.[24] Conversely, the function of the perfect to highlight the end-state of an action ought to increase the intensity of simulation directly pertaining to the endstates of described events while decreasing simulation of the nucleus of the event.

We tested these predictions in our lab[25] using the action-sentence compatibility effect paradigm, which by now will be very familiar to you. This is the type of study where people are shown to perform actions compatible with those implied by sentences faster than incompatible ones.[26] In line with previous work, we included sentences that described action toward or away from the body and had people respond by moving their hand toward or away from their bodies. The key innovation in our studies was to manipulate grammatical aspect. We had people read either progressive sentences or perfect sentences (examples in Figure 30). If what progressive aspect does for simulation is to augment mental focus on the ongoing action, then we should see a robust action-sentence compatibility effect with progressive sentences. And if perfect aspect really does shut off embodied simulation of the ongoing action itself, then we should see no action-sentence compatibility effect with perfect sentences.

	Progressive	**Perfect**
Away	*John is closing the drawer.*	*John has closed the drawer.*
Toward	*John is opening the drawer.*	*John has opened the drawer.*

FIGURE 30

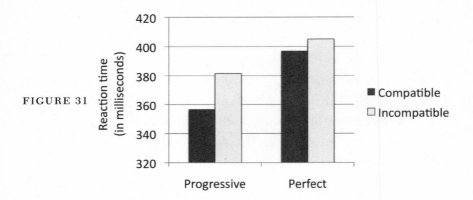

FIGURE 31

And that's what we found (results in Figure 31). The most plausible explanation for these results is that although progressive sentences drive understanders to mentally simulate the internal processes of described events, perfect sentences evoke less of this simulation.

What studies like these show is that whereas content words like *Mary, drawer,* or *open* point to specific experiences that can be simulated—like entities, states, events, or properties—grammatical aspect, like grammatical person, operates over the representations evoked by content words. Grammar appears to modulate what part of an evoked simulation someone is invited to focus on, the grain of detail with which the simulation is performed, or what perspective to perform that simulation from.

What Grammar Tells Us

One of the things that makes human language unique among animal communication systems is its highly developed use of grammar. As a result, figuring out how grammar works takes us a long way toward understanding what makes us—and our cognitive systems—unique. This chapter has laid out evidence that grammar contributes to embodied simulation in three ways. First, grammar combines and thereby constrains the contributions that individual words make to embodied simulation. (That's something that no other natural communication system we know of does.) Second, grammar can contribute meaning of its own. (Again, unlike other communication systems.) And third, it modulates embodied simulation, indicating not what to simulate, but how to simulate it. (Ditto.)

The idea that grammar modulates embodied simulation is enough to make a linguist's imagination run wild. What other pieces of grammar modulate embodied simulation? Does the fact that *barley* is singular while *oats* is plural make you simulate *a pile of oats* as being bigger or having more individual grains in it than *a pile of barley*?[27] If you're a native speaker of a language like Spanish, where all nouns have grammatical gender, do you mentally simulate a bridge (*un puente*) as having masculine characteristics but a key (*una llave*) as having feminine characteristics?[28] Is the difference between active *Mary punched John* and passive *John was punched by Mary* partly whose perspective you adopt? And the list goes on.

This last example brings up a critical and related issue. That's the issue of word order. One difference between *Mary punched John* and *John was punched by Mary* is the order in which *Mary* and *John* are mentioned. When you read or listen to language, you do so word by word, in order. This raises a question. Does the order of words affect the order of your embodied simulations? Do differences in word order across languages affect the order in which people simulate entities and events? And do you simulate as you go through a sentence, word by word, or do you hold off until all the words are in before launching into embodied simulation? In the next chapter, we'll take a little time to talk about this temporal dimension of simulation.

CHAPTER 6

Early and Often

Words and grammatical constructions are the building blocks that make up utterances—they're responsible for what utterances sound or look like and also for what and how you're going to simulate. But it isn't enough to know what the building blocks are. People actually using language also have to be able to recognize them in real time in an utterance and put them together. What are the processes that people actually use to deploy those building blocks? How exactly do you go from a sentence to a simulation?

At first blush, this might not seem like such a hard problem. After all, if you know all the words in a sentence and the grammatical constructions that put them together, then it's just a matter of binding all the pieces together and you're good to go. Like a puzzle, there's a single correct solution—only one way that the pieces will all click into place. Of course, with language it's not always that simple, because it's not always easy to tell exactly what was said or written. The very words someone uttered can be unclear (did he just say *can take* or *can't ache*?), or the signal can be degraded, for instance when you're talking on a phone or in a noisy environment. But even barring all that—even when you know exactly what words you're dealing with, it's still not easy to put all the pieces together, because of a very important and unavoidable characteristic of language: you don't get the whole utterance at the same time.

When you're reading or listening to language, the words come one after the other as they're spoken or as you run your eyes across the page. And they come at tremendous speed—around three or four per second in normal spoken language. This means that you have two choices, both suboptimal. The first is that you can wait until the end of the sentence and try to figure out at that point, when all the data has come in, what it was all about. The advantage to this strategy is that you have complete knowledge about the utterance when you make decisions. There are also disadvantages, though. The big one is that you're going to be slow on the uptake. If you wait until after the utterance to try to put it together, you're only starting to understand a sentence while the speaker is already on to the next one or waiting for you to reply or act or whatever it is that the utterance is trying to tell you. The second choice is to try to bind everything together incrementally: making decisions about what to do with each word, word by word, as you read or hear it. Incremental processing has the upside of allowing you to start understanding the utterance early, which, as long as nothing goes terribly wrong, means that you could potentially be keeping up with what the sentence is telling you as you go. And it has the additional advantage that if you're making best guesses about what the sentence is about as it unfurls, then as long as the sentence is pretty predictable, the things that come later are going to be less surprising and easier for you to integrate. But there's a big downside to incremental processing, and that's the fact that you're almost always dealing with incomplete knowledge. When you've heard just a few words of a sentence, you can guess where it's going, but when you're wrong, it can throw you for a loop.

So, because language is dynamic, fitting the pieces of an utterance together is somewhat harder than it first appears. So let's begin with the question of which option people go with. Are people relatively conservative, waiting until they have a complete picture of what the utterance is before trying to put it all together? Or do they jump on each word as it comes in, trying to incrementally fit it into an ongoing best guess of what the sentence is about? As we'll see, the answer is, perhaps surprisingly, that people do both.

Time Flies

How can we tell whether people process language incrementally? Well, if people are making early guesses about what the sentence is about, then that means that sentences that are relatively unpredictable should cause people problems, even over the course of the sentence, before it's over. It turns out there are such sentences—they're called *garden path sentences*, because they "lead you down a garden path" to the wrong outcome. For instance, here's a saying attributed to Groucho Marx:

> *Time flies like an arrow. Fruit flies like a banana.*

In reading the second sentence, *Fruit flies like a banana*, you probably mistakenly took *flies* to be the verb—after all, *flies* was a verb in the first sentence. You figured that *fruit* was the subject, and that you were going to find how it flies. But then you got to the *banana* at the end and realized that, no, *flies* wasn't a verb in this sentence but part of a compound noun—*fruit flies*. And *like* was actually the verb. You were duped because the second sentence didn't turn out how you expected it to.

Now, that example is a little messy, because there could be a lot of reasons why you expect *flies* to be the verb in the second sentence. For one, it could be that you expect the grammar of the second sentence to be like the grammar of the first. Or it could be that you just have a prior belief that *flies* is more likely to be a verb than a noun. But there are some cases where it's very clear what went wrong. For example:

> *The lawyer cross-examined by the prosecutor confessed.*

My guess is that, like most people, when you got to the *by the prose-cutor* part, you did a mental double take. Why is this? Well, the trick here is that lawyers tend to cross-examine people. So when you get to *cross-examined*, you think you've figured out the structure of the sentence. *The lawyer* is the subject, and *cross-examined* is a very apt verb for it to be the subject of. But then you got to *by the prosecutor* and had to go back and reconsider who did what to whom. And we know that part of the reason you were so easily led astray here is that lawyers tend to do cross-examining. To prove this to yourself, read the following sentence:

The witness cross-examined by the prosecutor confessed.

This should have been much easier to understand—you might not have gotten led down a garden path at all. Research shows that it would still be easier to understand, even if you hadn't just read another similar sentence.[1] The reason has to do with our knowledge about witnesses and lawyers. We know that cross-examining is the type of thing that lawyers tend to do, not to be subjected to, although the reverse is true for witnesses. This world knowledge affects our decisions about how to integrate words in the moment-by-moment activity of understanding the sentences they're part of. So it appears that the sentence understanding process uses aspects of meaning to come up with a word-by-word current best guess of what the sentence is about.

But how do we know, empirically, this is really happening in real time and not after the end of the sentence? One of the most productive ways to tell what's happening as people process sentences is to watch their eyes. Suppose you have people listen to sentences while looking at an array of pictures. The pictures depict various objects that differ in subtle ways. For instance, some of them are edible, like a cake and an apple, while others, like a cat and a table, are not. While they're looking at these pictures, the participants hear sentences that describe people acting on the objects. For instance, a sentence might be *The boy ate the apple*. When will people look where? Do people look more at edible objects when they come to the word *ate*, even before they hear the name of the object being eaten? When researchers in the United Kingdom asked this question, they found that people looked more at the set of objects that fit with the meaning of the verb, even before they heard the name of the relevant object.[2] For instance, they looked more at the apple and the cake than the cat and the table right when they heard *ate*, well before the end of the sentence. The most straightforward interpretation of this result, as well as subsequent work that has corroborated and extended it, is that people make predictions about what the rest of the sentence will contain as soon as words that they have already heard start to constrain what could reasonably follow. People start to cobble their understandings of sentences together incrementally.

Incremental Simulation

But when, you might well be wondering, does simulation kick in? In principle, there are again several possibilities. People might simulate early on in an utterance. Or maybe they are very stingy with simulation, waiting until all the pieces are in place—until the end of the utterance—to simulate whatever the combined story as a whole is supposed to be.

You might think that the second approach—the more conservative one—would be the more efficient way to go. It might help to do at least some parsing before you start to simulate. Before you can construct an embodied simulation to accurately depict *The rabid monkey is gnawing on the unconscious scientist*, for example, you need to know whether the word *rabid* applies to the monkey or the scientist, and the same for *unconscious*. You also need to know what the subject of *is gnawing* is—is the gnawer the monkey or the scientist? So it might make sense that before simulation of the described scene can take place, there must be a phase during which the understander deals with the grammar of the sentence to arrange its parts into sensible assemblies.

But on the other hand, couldn't it be that as we process words incrementally, we also evoke embodied simulation incrementally? That is, could it be that we mentally simulate as each word comes in, while we're still trying to parse a sentence into a coherent whole? Let's apply this idea to the garden path sentences we saw earlier, *The lawyer cross-examined by the prosecutor confessed*. When you hear *the lawyer*, you might conjure up a simulation of a lawyer, though perhaps an unspecific one. Then comes the word *cross-examined*. You've got your mental representation of a lawyer, and, although the current parse you have for the sentence is incomplete (and as it will later turn out, incorrect!), you go ahead and attempt to mentally simulate the lawyer, standing in front of a witness stand, cross-examining. Whom the lawyer is cross-examining is unclear, so your embodied simulation might well omit details of what the cross-examinee looks like, or you might simulate a prototypical or stereotypical cross-examinee. But when you then get to *by the prosecutor*, your parsing machinery grinds to a halt, and so does your simulation machinery. And so, once you've collected enough information to reparse the sentence as it was intended—the lawyer is in fact the cross-examinee—you can resimulate the described scene, this time with the lawyer on the witness stand, rather than in front of it.

So do you simulate early and often or late and infrequently?

To answer this question, we again need to find a way to detect embodied simulation over the course of sentence processing. In fact, there's a very widely used methodology known as *self-paced reading* that allows us to do just this. There are different variants, but the basic idea is that you display parts of a sentence on a screen one at a time. The catch is that the parts don't appear automatically—you have the person reading the sentence perform some action to make each piece appear on the screen. Typically, self-paced reading uses a button, so a participant in an experiment might push the button and see *The* and then push the button again to see *rabid* and again to see *monkey* and so on. Self-paced reading is often used to detect where people have trouble processing sentences. Logically, the more difficult it is for a

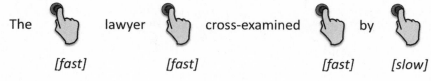

The   lawyer   cross-examined   by

[fast]   *[fast]*   *[fast]*   *[slow]*

FIGURE 32

person to understand a given word that has just appeared on the screen, the more time they'll need to understand it, and, as a result, the longer they'll take to press the button to display the next word.

Self-paced reading can also be adapted to tell us about embodied simulation. Researchers at the University of Rotterdam used a device we discussed earlier, the Knob, which, as you might recall, is a very small device—one inch in diameter—that you turn to respond to stimuli.[3] The Knob can be configured so that in order to see the next part of a sentence you're reading, you rotate the Knob a very small amount—just five degrees. The neat thing about the Knob is that you can have people rotate it either clockwise or counterclockwise. What the Dutch researchers did was to have people use the Knob to pace their own reading of sentences that described manual rotation either clockwise or counterclockwise. For instance, the two sentences below describe actions involving rotation of the dominant hand in opposite directions. (The slashes you see show the segments that were presented with each five-degree rotation of the Knob.)

> *Before / the / big race / the driver / took out / his key / and / started / the / car.*
> *To quench / his / thirst / the / marathon / runner / eagerly / opened / the / water bottle.*

With sentences describing clockwise or counterclockwise motion and responses involving one of the same two types of motion, you can measure whether there are compatibility effects between the sentence meaning and the manual response direction. This is a lot like the action-sentence compatibility effect, except that measurements are taken

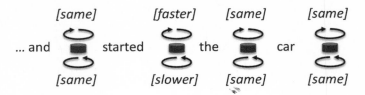

[same]   [faster]   [same]   [same]

... and ⟳ started ⟳ the ⟳ car ⟳

[same]   [slower]   [same]   [same]

FIGURE 33

not after the sentence is over but right in the thick of it. The findings
from this study were quite clear. When people were rotating the Knob
in a compatible direction, they turned the Knob faster. For instance, in
the example above, if they were rotating the Knob clockwise, this was
compatible with the described action of turning a key in the ignition to
start a car, and so they responded faster after the word *started*. But,
when they were turning the Knob in an incompatible direction, they
were slower. Crucially, as you can see in Figure 33, this was only true
of the rotation actions the readers made directly after the verb (for instance, *start*). When they saw the words before or after the verb, they
showed no difference, regardless of whether the Knob was being
turned in the same direction as the object in the sentence.

The graph in Figure 34 depicts their results more clearly. What you
see there is the average time it took people to rotate the Knob (either
clockwise or counterclockwise) while processing different regions of a
sentence. The two lines depict cases where the direction of rotation implied by the sentence matches or mismatches the direction of the action
the participant makes to see the next part of the sentence. As you can
see, there's basically no difference at any of the regions, except in the
Verb region. If you look at the sentences on the previous page, you can
see why this would be. The relevant verbs there are *started* and *opened*.
Neither of them necessarily describes rotation in a particular direction
on its own. But the stuff before the verb in each sentence sets up the
context such that, when the verb appears, it's pretty clear that the sentence describes rotation in a particular direction.

So the big takeaway point here is that as soon as a verb appears
that implies manual rotation in a particular direction, people mentally

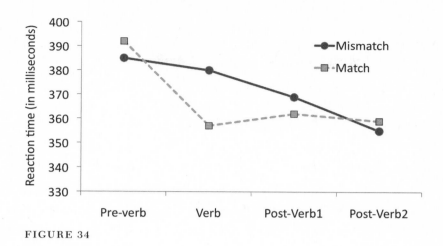

FIGURE 34

simulate rotation in that direction. This answers our question about incrementality; people don't wait until the end of a sentence to start simulating—they do it early on, in the middle of the utterance.

But this result is actually quite revealing for three other reasons as well. First, this effect is produced by words like *open* and *start* that don't in and of themselves denote rotation in a particular direction; they only have these meanings in the right context. And yet it's when people see these particular words that the evidence for embodied simulation of hand rotation shows up. So it seems like it's not what the words mean in isolation that guides people in their incremental simulation. Instead, people are calculating what the words describe in the particular context of the sentence they appear in, and use this to run embodied simulations.

A second key point is that the place where embodied simulation effects showed up in this study is right at the point in the sentence where it became clear that the sentence was about rotation in one direction or the other. If so, this suggests one way in which simulation is less than completely incremental. Do people simulate as early as possible, whenever the sentence gives them reliable information about something to be simulated? We get a clear answer from some additional work conducted with the Knob by the same researchers. They conducted another study, to see whether words other than verbs can produce the embodied simulation effects they observed.[4] So they created sentences that didn't imply rotation in one direction or the other until well after the verb—some examples are below. As you can see, it's only on the very last word of the sentence that you can infer which direction the carpenter had to rotate his or her hand to turn the screw. And sure enough, the researchers found the same match-mismatch difference, but this time it only appeared on this last word of the sentence and not on the words preceding it, including verbs like *turn*. So it appears that whenever in a sentence you get distinguishing information about a simulation, you go ahead and mentally simulate, whether the key word is a verb like *started* that describes the action, or an adverb like *loosely* that's only, well, loosely associated with it.

> *The carpenter / turned / the / screw. / The boards / had / been / connected / too / tightly.*
> *The carpenter / turned / the / screw. / The boards / had / been / connected / too / loosely.*

A final important outcome of this study is the finding that the simulation effect is ephemeral—by the time the reader is on to the next part of the sentence ("Post-Verb1" in the graph above), there's no measurable difference between the matching and mismatching conditions. This last point is worth thinking a little more about. Why would it be that the measurable simulation of rotating your hand in a particular direction would disappear immediately after the verb?

Here's one possibility. If you look at the sentences about starting cars and opening water bottles, you'll see that the part of each sentence immediately after the verb indicates the thing that the action was performed on. At this point, people reading the sentences don't appear to be representing manual rotation anymore. Why not? Maybe what's happening is that with each new word that comes in, you shift your attention to whatever the new information is that it's providing, and construct the appropriate embodied simulation. For instance, in the sentences above, the first word after the verb is *the*. Maybe what *the* tells you is to prepare to mentally simulate some object, perhaps motor aspects of how you interact with it, or other aspects, like its visual appearance, tactile feel, and so on. By contrast, if instead of *the*, the next word were something related to the manual rotation described by the verb, like *slowly*, perhaps you would be more likely to continue simulating the rotation as you processed it. As it turns out, the Knob has something to tell us about this, too.

What the Knob's handlers did was to come up with new sentences to present to readers, who once again knobbed their way through the sentences. The sentences were similar to those in the previous experiment, except that they now had a specific word after the verb, an adverb. There were two types of adverb, as shown in the sentences below. One type described the manner of the action (such as *rapidly*, *slowly*, or *quickly*), as in the first sentence below. The other described an aspect of the mental state of the person performing the action (like *hungrily*, *obediently*, or *happily*).

> *He was / craving a / juicy / pickle. / On the / shelf, he / found a / closed jar / which he / opened / rapidly.*
> *He was / craving a / juicy / pickle. / On the / shelf, he / found a / closed jar / which he / opened / hungrily.*

They figured that if the adverb was about the person's mental state and not the action itself, then this would draw readers' embodied simulation away from the rotation. But if the adverb was about the action, then they should continue simulating the action. What they found was that adverbs that described the manner of action, like *rapidly*, sustained focus of the embodied simulation on the action itself even after the verb—people were faster to perform matching than mismatching actions on both the verb and the subsequent adverb. But with adverbs like *hungrily*, which pertain to the emotions or mental states of the person performing the described action, people showed a significant match-mismatch difference only for the verb and not the subsequent adverb. In other words, how long you keep simulating action depends on whether the following words draw your attention away from that aspect of the described scene.

So the Knob studies tell us a lot about incremental simulation. People activate simulation incrementally. They do so in a context-sensitive way; it only shows up when a word in its particular context implies something to simulate. And they do so opportunistically—as soon as readers have reliable information to simulate, they do so, regardless of whether the word that finally cemented it is a verb or an adverb or whatever. And finally, this simulation is ephemeral. It disappears quickly when a sentence transitions to different content to be simulated. This evidence is enlightening and compelling. But again, we'd be most comfortable with these conclusions if we found convergent evidence using a different method.

Fortunately, a different strand of research[5] addresses the temporal dynamics of simulation as well, even though it started out focusing on a very different issue. This work came out of my lab. We wanted to know whether embodied simulation works the same in other languages as it does in English. A graduate student, Manami Sato, was doing research on Japanese, and she tried to replicate one of the seminal experiments on visual simulation. This was the one where people read sentences that mention objects and imply that they have certain shapes. For instance, each of the sentences below implies a different shape for the egg, namely the shape to its right.

Nana put the egg in the fridge.

Nana put the egg in the pan.

FIGURE 35

The original finding using English sentences was that people were faster to press a button saying that the depicted object had been mentioned in the sentence when the shape in the picture matched the one implied by the sentence.[6] We tried to replicate this finding with Japanese sentences, expecting to find the same thing—there's no reason to think that Japanese should be less coercive in driving simulation of object shape than English is. But we were surprised to find that there was no difference in participants' reaction times, regardless of whether the picture of the object matched the one implied by the sentence. At first we were baffled, so we looked closely at the sentences we used as stimuli in Japanese. They looked like this:

Nana-ga	*reezooko-nonakani*	*tamago-o*	*ireta*
Nana	fridge-inside	egg	put

"Nana put the egg inside the fridge"

Nana-ga	*huraipan-nonakani*	*tamago-o*	*ireta*
Nana	pan-inside	egg	put

"Nana put the egg in the pan"

Then it struck us that perhaps we were testing for embodied simulation at the wrong time. After all, Japanese sentences have a different word order than English ones. In English, the verb (like *put*) usually comes in the middle of a sentence, and the direct object whose shape we were testing for (like *egg*) comes later. So if you probe simulation (using a picture, as this method does) at the end of the sentence, you're getting at people's simulations right after they've discovered what shape the object should have. But in Japanese, the order is reversed, as you can see from the sentences above. In Japanese, you build up an idea of what objects are where and what shape they have pretty early in the sentence. By the time you read the word *tamago*, meaning "egg," you have a pretty good idea what shape it has—you've already seen that it's "in the pan" or "in the refrigerator," and eggs don't tend to be placed into pans intact any more often than they're cracked open into refrigerators. So when the Japanese reader finally gets to the verb *ireta*, meaning "put," at the very end of the sentence, there's no need to perform any more visual simulation of the shape of the object. Instead, this verb might trigger simulation of something else. For instance, it might yield motor or visual simulation of the action of putting the egg in the described location.

We wanted to test this proposed explanation of the otherwise mysterious difference between English and Japanese. So we made a little modification to the methodology, changing where in the sentence the picture appeared. If our explanation was right, then moving the picture earlier in the sentence—so that instead of coming at the end, it directly followed the object noun (e.g., the word *tamago*, meaning "egg")—this would lead to a difference between the matching and the mismatching pictures. So we conducted another experiment, using the same kind of sentences, but this time people saw the picture before the verb and had to decide whether it had been mentioned in the sentence. So, for instance, they would see:

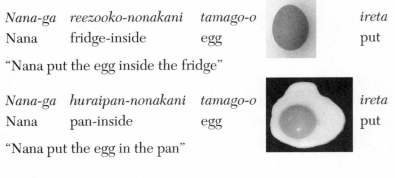

| *Nana-ga* | *reezooko-nonakani* | *tamago-o* | | *ireta* |
| Nana | fridge-inside | egg | | put |

"Nana put the egg inside the fridge"

| *Nana-ga* | *huraipan-nonakani* | *tamago-o* | | *ireta* |
| Nana | pan-inside | egg | | put |

"Nana put the egg in the pan"

FIGURE 36

When we looked at the results, all of a sudden, the mysteriously missing effect was back—Japanese speakers responded faster to the images when they matched the shape implied by the sentence than when they didn't. So we can conclude several things from this. As the Knob showed, people simulate details (in this case object shape) even before getting to the end of a sentence. They do so in not a word-specific way but use the word plus the context (the word *tamago* "egg" was the same in both conditions). And that simulation is short-lived, because any sign of it disappears by the end of the sentence.

The implication of these results is that on occasion you might mentally simulate one thing because you're pretty certain that you know what's being described, only to have to revise your simulation entirely when some new word appears later in the sentence. In other words, maybe you sometimes get led down a simulation garden path. For example, imagine you're a Japanese speaker, and you read:

Nana-ga	*reezooko-nonakani*	*tamago-o*	*otoshita*
Nana	fridge-inside	egg	dropped

"Nana dropped the egg inside the fridge"

or

Nana-ga	*huraipan-nouede*	*tamago-o*	*korogashita*
Nana	pan-on	egg	rolled

"Nana rolled the egg on the pan"

These sentences imply unusual shapes for the objects in the mentioned locations—you expect an egg in a pan to be broken, not capable of rolling. But because in Japanese the verb comes at the end of the sentence, you don't actually know until the very last word that the shape of the object is the unexpected one—that the egg in the fridge is actually broken, or that the egg in the pan is actually intact. We already know from the experiments above that prior to the verb, people have constructed embodied simulations of the shapes of objects. In order to understand sentences like *Nana rolled the egg on the pan*, when the verb appears at the end, people might actually have to construct an alternative simulation of the object, one with a different shape.

To test this, we used sentences like these, which implied one shape up to a point and then switched it at the end and presented the picture at the end of each sentence. We thought that if people were really switching back and forth from one simulation to the other, then they should be faster to respond to an image matching the final shape implied by the unexpected verb. And sure enough, when we looked at the results, the match-mismatch effect was back. People responded faster to the shape depicting the final, implied shape of the object than to the one that would have seemed more reasonable just one word before. In other words, people quickly correct and revise the embodied simulations they construct when new information overrides the assumptions they had made previously and that they had based the previous simulation on.

In a Word

So people process sentences incrementally. But let's zoom down for a second to a smaller scale. When exactly does the simulation start? We already know that a word in the right context can lead people to simulate. But do people wait until the word is over, or do they start making guesses early on, with only partial information? This is a question that operates over an extremely tight timescale—a monosyllabic word often lasts no longer than a couple hundred milliseconds. So we need a tool that allows us to make inferences about what people are simulating in real time, and a very useful tool for this purpose is again eye tracking.

As we saw earlier, we can track how people's eyes move to images in their environment while they're listening to words and sentences and use what they look at, and when, as an index of what they're thinking, and when. When it comes to words, people don't wait a tenth of a second to look at what they think the word is about. Suppose you have someone looking at pictures of a number of different objects, say a beetle, a beaker, a speaker, and a carriage, like in Figure 37.[7] If you have that person listen to a sentence about one of the objects, like *Pick up the beaker*, while you're tracking their eye movements, you'll see that they look more at the beaker than the carriage or the speaker. That might not be surprising—after all, the person is being told to interact with the beaker, so as long as they know what a beaker looks

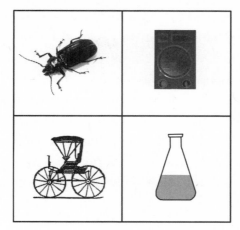

FIGURE 37

like, they should look at it more than other objects. The interesting thing, though, is the timing of that increased looking. Just after the beginning of the word, even before the *k* in *beaker*, people are already starting to make increased eye movements to the breaker.

What's more, the clever thing about this particular experimental setup is that the first couple sounds of *beaker* (namely, the *b* and the *e*) are ambiguous between *beaker* and *beetle*, and, as a result, people actually look more at the beetle early on, too, at least right after they've heard the beginning of the word *beaker*. Then, as soon as they hear the *k*, they stop looking at the beetle. That means that they're making guesses as soon as they start hearing a word about what in the environment it could possibly refer to.

But when people start to look early on in a word at things in the world that they think it refers to, how do they know what to look at? For a word that starts *be* . . . , how do they know that it's the pictures on the top left or bottom right that are most likely to be the thing they're hearing about? Most obviously, they have to be predicting what the rest of the word will be, starting with just the first couple of sounds. This by itself is quite a task. But there's a deeper issue beyond that. Assume you've predicted that the word you're hearing is going to be *beaker*. How do you know which image depicts something that could be called by that name?

Presumably, people do this by accessing some kind of perceptual knowledge—knowledge about what beakers and beetles look like. And we can actually detect whether this is what's going on by looking at the kinds of mistakes people make in a task like the one with the beaker and the beetle but modified slightly. Suppose you hear a sentence that mentions a *frog*, but there isn't a frog on the screen in front of you. There is, however, a piece of lettuce, along with a couple other things . . . a guitar, a rope, a strawberry. In this case, all other things being equal, eye-tracking shows that you're most likely to look at the lettuce, more so than at other objects.[8] But, if instead you hear a sentence about a snake, then you're most likely to look at the rope.[9] Why the lettuce in one case and the rope in the other? As you may have surmised, the remarkable finding is that people tend to look at things that look similar to the thing that was mentioned, either in terms of the color, as in the case of the frog and the lettuce, or their shape, as in the case of the snake and the rope. This is important, because it means that while you're hearing a word, you're already activating the perceptual details of the things you think it refers to.

And Then?

At this point, you might have noticed that some of the results we've reviewed seem to contradict one another. On the one hand, we've seen again and again that, at the end of sentences, people act as though they're mentally simulating. For instance, the very first Knob study, described back in Chapter 4, showed that when people make judgments about the meaningfulness of sentences after reading them by twisting the Knob one way or the other, they do so faster when the action described by the sentence is in the same direction. But, on the other hand, in this chapter, we've seen that simulation is incremental and ephemeral. Although you conjure up embodied simulations of elements mentioned in a sentence as you go, this simulation is soon obscured by simulation effects produced by the next informative piece of language in the sentence. If a simulation that emerges during a sentence subsequently disappears when you get to the next word, then why do we see simulation effects at the ends of sentences?

The short answer is that we don't know. I mean, we know that when the last word in a sentence provides key information that leads to a viable simulation, people are going to be simulating that detail at sentence's end. But we don't know why several seconds after the end of a sentence, some simulation effects show up again, even ones that have ostensibly flared up and then been extinguished over the course of the sentence. But just because we don't know doesn't mean we're at a complete loss for ideas. For example, one relatively popular hypothesis is that you go through several steps during sentence processing. There's a first stage where you build up the pieces of the sentence from what you hear or read. And then there's a second stage where you resimulate what it was that you were supposed to simulate.[10] This second stage is often referred to as *wrap-up*[11] because you're wrapping up what the sentence was ultimately about. So wrap-up could be why we see simulation effects both during a sentence and after the sentence. Early, incremental simulation is driven by the current and immediately preceding language, while wrap-up is a slower process that reruns some key parts of the incremental simulation.

What would lead people to perform wrap-up simulations? Perhaps they serve a purpose. You appear to go through a lot of twists and turns in simulating as you read or listen to a sentence, but sentences often convey overriding ideas that you don't want to jumble up. So once you're certain you've come to a viable interpretation of a sentence, perhaps a recap is in order. You could use wrap-up for a number of more specific purposes. For one, a wrap-up simulation might be something relatively reliable you can use to update your beliefs about the world, whereas you might not want to rely on the ephemeral simulations that you experience in the middle of a sentence for this purpose. For the same reason, you might prefer to use wrap-up (over the short-lived, ephemeral, and often ultimately incorrect simulations produced during a sentence) when preparing to act in response to an utterance. The appropriate response generated in the middle of a sentence like *There's a bomb in my office . . .* might not seem so appropriate when you get to the end of the sentence . . . *clip-art collection.*

Wrap-up simulations might be particularly well suited for these purposes, and not just because they're more likely to be accurate. They can also in principle last longer than simulations performed during the course of sentence comprehension, because they're less likely to updated or changed on the basis of subsequent input. Holding a stable and (likely) accurate simulation firmly in mind for a while could be quite useful if you want to use your understanding of an utterance for some purpose.

But for now, the role of wrap-up is at best speculation. The relation between what you do while you're processing a sentence and the embodied simulations you perform subsequent to it is an area ripe for future study.

Say It Ain't So

But the process of understanding a sentence doesn't necessarily end with wrap-up. You've probably had the experience of coming to the end of a sentence, only to have to do more mental work to figure out what the sentence actually means. For instance, suppose you're hiring, and you're scanning through a letter of recommendation for a particular candidate, when you come upon a sentence like this: *I can't recommend this candidate any more strongly.* Your first impression might be that the recommender is offering their strongest support for the candidate. That is, that she can't recommend the candidate any more strongly because there are no words strong enough to express how unfathomably amazing the candidate is. But upon further reflection, you might be struck by the possibility that the recommender is actually saying that she can't recommend the candidate any more strongly because she doesn't know much about the candidate or that there's nothing good to say about him. The point here is that the process of understanding a sentence often continues, dynamically, even after the last word is heard or the last punctuation mark read.

Nowhere is this more obvious than in negated language. So far, we've focused only on language describing purportedly factual scenes. But every language in the world also provides means to describe events that didn't happen, won't happen, or aren't happening. English carries several sorts of negation in its grammatical arsenal. An entire sentence can be negated, like so: *The shortstop didn't hurl the softball at the first baseman*. Parts of a sentence can also be individually negated: *No shortstop hurled the softball at the first baseman* or *The shortstop hurled not one softball at the first baseman*, and so on.

Negation is of special interest to accounts of language understanding that use simulation, because it is a particularly challenging case. Even if there is convincing evidence that people perform embodied simulation when understanding language about things happening, like softball players throwing things at each other, it's not immediately obvious how the same process would apply to language about things <u>not</u> happening, like softball players <u>not</u> throwing things at each other. What—if anything—do people mentally simulate when confronted with negated sentences?

One leading idea is that negation is processed through several stages. When you hear or read a negated utterance, the first thing you do is to mentally simulate the *counterfactual* scene—that is, the scene described as not true. So, for instance, in understanding *Your birthday presents aren't on top of the refrigerator*, the counterfactual scene is the (false) one where presents are in fact on top of the refrigerator. But this counterfactual simulation is subsequently suppressed or modified, which leads to the second proposed stage in the course of processing. This is the activation of the *factual* scene—the scene that the sentence is saying is actually true. In our example, this would mean a simulation of the presents not on top of the refrigerator, but somewhere else, as you can see represented schematically in Figure 38.

time →

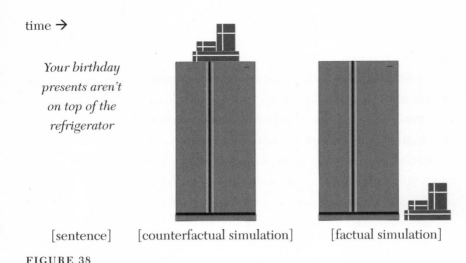

*Your birthday
presents aren't
on top of the
refrigerator*

[sentence]   [counterfactual simulation]   [factual simulation]

FIGURE 38

So let's first look at whether this is actually what people do, and then we can try to understand how it would allow people to understand negated language.

So, how can we tell—empirically—whether people are first performing a counterfactual then a factual simulation? The simplest approach would be to use the familiar methods to assess simulation but to present the critical stimuli at different times in processing. Logically, we should see effects of the counterfactual simulation being activated first, then the factual one. We'll start with some studies that address what happens immediately after the sentence and then turn to what happens in the seconds following it. Work by German psychologist Barbara Kaup and her colleagues used a method very similar to the one described in the study on Japanese to look at what happens right after a sentence. They first presented people with negated sentences, like *There was not an eagle in the sky* or *There was not an eagle in the nest*, which mentioned entities (the eagle, in this example) that could take on different shapes.[12] Then, 250 milliseconds later, the experiment software displayed a picture depicting that same entity in one of the two possible shapes. The question was which image people would process faster. With affirmative sentences,

we've already seen the outcome—*There was an eagle in the sky* produces faster responses to the flying than the sitting eagle. What about
with negated sentences?

If people first simulate the counterfactual scene and then the factual one, then at this very early time (just a quarter of a second after
the sentence disappears), we would predict faster responses to images matching the shape of the object implied in the sentence's counterfactual situation—that is, of the flying eagle in this case—than
when they did not. And this is what Kaup and her colleagues found. If
the sentence tells you that there was not an eagle in the sky, you nevertheless simulate a flying eagle.

There's other evidence pointing to this same conclusion that comes
from a different methodology, a method called *semantic priming*.[13]
Rachel Giora and her colleagues had participants first read a sentence
of one of three types (all of which are below): an affirmative sentence
containing a priming word (like *sharp* below); a negated version of
the same sentence; or an antonym sentence—an affirmative sentence
using an antonym of the priming word (like *blunt*). (These are actually
translations of the original sentences, which were all presented in Hebrew to Hebrew native speakers.)

Affirmative	*This instrument is sharp.*
Negated	*This instrument is not sharp.*
Antonym	*This instrument is blunt.*

People were supposed to indicate they had understood each sentence
by pressing a button, and 100 milliseconds later, they saw a sequence of
characters that either formed a word or did not. Their job was to decide
if this sequence of characters was an actual word in their language, so
they were performing what's called a *lexical decision* task. In the critical
case, the target word was a real word that was either related to the priming word (e.g., *piercing*) or unrelated to it (e.g., *glowing*). What Giora
and her colleagues found was that affirmative sentences lead to faster
responses to related prime words than unrelated prime words. This isn't

surprising—we know from lots of studies on semantic priming that words are processed faster after other words that have related meanings. Interestingly, though, the negated sentences did the same thing, suggesting that even when a word appears in a negated sentence context, it still activates mental representations of its meaning. And what's more, we know that this priming in negated sentences is not due to the factual situation (where the instrument is blunt) being activated, because in the antonym condition, the related words were not responded to any faster than the unrelated words were. That is, just activating the idea of bluntness does not prime the word *piercing*, but *not sharp* <u>does</u> prime the word *piercing*. This suggests that the effect in the negated condition did not result from activation of the factual situation but of the counterfactual situation.

Now, this finding alone doesn't necessarily mean that people are performing embodied simulations when they process negated sentences. It could just be that the word *sharp* makes you respond faster to the word *piercing* because their meanings are related, regardless of what embodied simulation you do or don't perform. And you would be quite justified in thinking this. But it does at least overlap with the previous study—the one with the eagle not in the sky—in showing that in negated sentences, the counterfactual situation is indeed accessed right after the sentence is over.

So these very different studies both provide evidence in favor of at least the first part the hypothesized two-step process. Is the factual situation mentally simulated next? To know, we'd need to probe people's embodied simulation a little later on. The two studies we just looked at presented the image or word 250 and 100 milliseconds after the end of the sentences, respectively. But if you wanted to know what was happening farther down the line, you might present the image 750 or 1500 milliseconds after the end of the sentence. This is what Kaup and her colleagues did.[14]

They created sets of sentences, like the ones in Figure 39, which again mentioned objects that could be in different states. They again had people read a sentence then push a button to indicate they un-

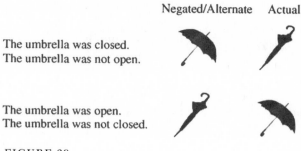

FIGURE 39

derstood it. This button press then triggered the presentation of an image on the screen—of the object in one of those two states. But they introduced a delay of 750 or 1500 milliseconds before each picture. What they found was exactly in keeping with the two-step account. In Figure 40, you see how long it took people to respond to the picture. On the far left is the outcome from Kaup's original study, where the delay before the image appeared was 250 milliseconds. As you recall, people responded faster to the counterfactual image. But at the 750 millisecond delay, there was no difference between the factual and counterfactual images. And by 1500 milliseconds, the trend had reversed itself—people were now faster to say that the factual image had been mentioned than the counterfactual one. This is pretty

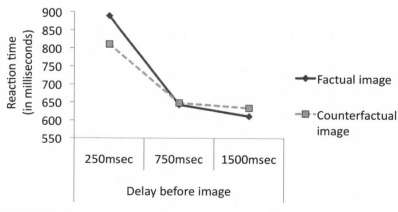

FIGURE 40

compelling evidence that when you process a negated sentence, you first construct a simulation of the counterfactual scene and then move on to the factual one.

This seems like a tremendous amount of work to do just to understand a sentence. What's it all good for? It's possible that each part of the process serves specific functions. The factual component seems like it ought to be the most important, because in a way, it's like a wrap-up simulation—it encodes the critical information that would allow the understander to act appropriately. If someone tells you that *Your birthday presents aren't on top of the refrigerator* (and if you believe them) you shouldn't expect it to be on the refrigerator and shouldn't look there for it. You might instead look for it elsewhere or ask the speaker where it is.

But if it it's correct that the factual simulation does this work, what good is the counterfactual simulation? First, it could be part of the process that allows you to determine what the factual situation actually is in the first place. By phasing out and suppressing your simulation of the counterfactual scene (for instance, the present atop the fridge), you can ensure that the factual simulation that you end up constructing is inconsistent with what the speaker intended. Namely, it has to not have whatever the negated properties of the counterfactual scene are. A second possible function of the counterfactual simulation is that by going through the counterfactual phase, you might actually take a particular type of trajectory in understanding an utterance that is different from what you would do if you just heard an affirmative sentence describing the same factual scene. Compare the *umbrella isn't open* with *the umbrella is closed*. You might end up with the same factual simulation in processing these two sentences (of a closed umbrella), but you will have taken different paths to get there. And just as it makes a difference whether you cheated your way to a 2400 on the SATs (yes, it's true—the SATs are now scored out of 2400) or studied assiduously for months to produce the same outcome, so it might make a difference what cognitive path you traverse to get to a factual simulation. The experience you have of simulating and then rejecting a counterfactual simulation might be part of what it is to feel like you understand a negated utterance.

Meaning on the Run

There's actually something quite profound about this idea—that the process of understanding might be as important as the outcome. And you can see this not just in negated sentences, but pervasively throughout language. Take an example sentence like this:

My sister was a track star.

You can say it in different ways, with different intonation, to make different points. (Intonation is the tune, emphasis, and speed of an utterance.) For instance, if you're explaining how it shouldn't be surprising that your sister miraculously just went out and ran a marathon, right after giving birth to her second child, you might put emphasis on *track star*, so the sentence sounds like this:

My sister was a TRACK STAR.

But you can say the sentence in other ways. Suppose your sister was indeed a track star in the past but has since taken to sitting on the couch and watching daytime television while eating bonbons. Then if you want to assert that your sister, now rotund and immobile, was indeed at one point a successful athlete, then you might say:

My sister WAS a track star.

What's the difference in how people will understand these two versions of the same sentence? One idea is that when you emphasize *was*, it tells the listener that what the sentence describes used to be true but isn't anymore. A graduate student working with me, Heeyeon Yoon, investigated this in her dissertation research, and it appears that sentences like this one, with emphasis on the *was*, act a lot like negated sentences, in that they drive you to go through several phases of simulation. Instead of a counterfactual and then a factual simulation, though, sentences like *My sister WAS a track star* lead you through a simulation of the formerly true scenario (the sister being a track star), followed by a simulation of the implied, different one (her no longer being a track star). This contrasts with the one-stage process you appear to use when understanding the more neutral *My sister was a TRACK STAR*, where you simulate only the sister as a track star. With this latter intonation, there's no implication that the sister has stopped being a track star, so no need for a modified second phase of simulation.

So subtly different sentences could result in subtly different courses of simulation. For another example, consider hyperbole. Suppose your teenage son comes home and tells you:

I'm so hungry I could eat a goat.

He probably doesn't really mean that he wants you to prepare him a meal made of one entire adult goat, or even a meal of equivalent caloric value. No, he just wants you to infer that he's very hungry. And that's probably the conclusion you come to. But how? Well, again, it could be that you pass through a phase in which you do simulate the described, but unrealistic, scene of your son going to town on a goat carcass, after which point you come to the more sensible conclusion that your son is hungry, and you then perform an appropriately better-mannered simulation.

And it's not just hyperbole; many other sorts of figurative language offer themselves as excellent candidates for multiphase simulation. One example is metonymy, which we'll look at in Chapter 9. This is where you use a word that denotes something to refer to something else related to it. For instance, you might refer to a job candidate as *red hat* because he was wearing a red hat at the interview. When you use metonymy like this, you are probably leading people through a process of simulation that first evokes a simulation of the red hat, followed by the rest of the person underneath it. If you call him something else, say *freckles,* then your simulation will follow a different course. Even though *red hat* and *freckles* both lead you to identify the same person, the course of simulation that you traverse to get there might be quite different.

This dynamicity might be quite important for understanding how meaning works. (Note that we're stepping away here from what the experimental research itself says, and into a largely more reflective mode. So take what I have to say here with a grain of salt. Not literal salt.) As we've seen in this chapter, understanding is dynamic, in at least two ways. First, as you hear or read words and parse them incrementally, you construct embodied simulations of your current best guess of what they contribute to simulation. And, second, some language makes you go through a series of steps during simulation.

But all of this is quite different from how we typically think of meaning. Usually, when we want to know about meaning, we ask things like *what does X mean?* and expect the answer to be a description of a thing or an event or an idea, as if there were a description that could capture what a word or phrase or utterance means. I suspect that's how most people who have never formally studied meaning see things. (Though I admit it's hard for me to remember what that was like.)[15] But it's certainly the case that professionals who develop technical theories of meaning almost to a person make the same assumption. Linguists try to come up with the categories or logical symbols or descriptions that can aptly describe what it is that words and utterances mean. Psychologists and computer scientists try to come up with simple, elegant accounts of how to characterize the building blocks of meaning and how they combine.

But the thing or event or idea that language refers to isn't the whole story. The way that you understand *freckles* and *red hat*— the processes that you go through to get from the words to a simulation of the person referred to—are different. It's true that you can come to the same endstate in the different ways. But you have different experiences of getting there. This means that your subjective experience will be different—you might for example experience simulations of the guy having the different properties that he's being referred to in terms of. You might be more likely later to remember him having those properties. And you might make different inferences based on those properties (like about his race, or the weather, or what college he went to).

In other words, maybe *what does X mean?* is the wrong question, or at best maybe it's only part of the question. Perhaps the real question is *what are the understanding processes that X invokes?* How you put the pieces together and what steps you go through once you've assembled them may be equally important as what the pieces are like themselves. Think about a sentence as a jigsaw puzzle. If the picture on the front were the only interesting part about a puzzle, then you'd never bother to open the box in the first place.

CHAPTER 7

What Do
Hockey Players Know?

Thus far we've emphasized the features of embodied simulation that people have in common. But the truth is far more nuanced. People don't all simulate the same things or simulate them in the same way. For one thing, the embodied simulations you construct are based on the experiences you've had. If everyone had the same experiences, they'd be able to run the same simulations. But they haven't, so they don't. Even something as banal as the word *dog* has qualitatively different effects on people. For someone who was bitten by a dog as a child, the mere mention of the word *dog* can evoke images of a massive, snarling, ferocious beast with frightening incisors—images that the word *dog* doesn't usually trigger for the rest of us, except upon extended reflection or in the right context. Our unique personal histories produce stark differences in everything from how we reason to what we value, and so it shouldn't be surprising that they also affect how we understand language.

There's a second way that embodied simulation differs across people. As you've surely noticed, we don't all have precisely the same cognitive capacities or cognitive preferences. For example, I have a really bad visual memory. I honestly can't tell you what color my bathroom is. I'm pretty sure it's either pink or blue. It might be green. I'd have to check. But on the other hand I can often remember the exact words someone said, verbatim, a week after hearing them. Other people's minds work differently. And as a result, people have different resources at their disposal to deploy when understanding language. Someone with a really reliable, detailed visual memory might use their vision system more in understanding, as compared to someone like me, who might use their motor or auditory system more.

How much do these differences matter? If simulation is really used for comprehension, then you might think that a speaker and a listener need to run very similar simulations to understand each other. And it's quite possible that compatibility or incompatibility of embodied simulations can make it easier or harder to communicate with people. To the extent that we have similar mental lives, due to similar personal histories or cognitive abilities, communication can feel effortless. When our mental lives differ, communication can grind to a standstill. But, at the same time, people are still often able to make themselves understood, even when simulating differently from the people they're communicating with.

Experience Matters

Like everyone else, you are an expert in something. It might be knitting, theoretical astrophysics, car repair, or National League baseball. But there's something you know more about than the average bear. And this expertise is the product of experience: lots and lots of experience. As an expert, you know things that the rest of us do not. So the question is, Do you also understand language about the particular domain you're an expert in differently from novices? For instance, an expert handyman might easily and quickly conjure up a precise and accurate mental representation of a crescent wrench, while someone less experienced might confuse it with a monkey wrench or not be too sure whether or not it has moving parts (it does). So communication between an expert and a novice could quickly break down. Consider the following interaction between Serge, who's a thoracic surgeon, and Pete, who was just making a pizza delivery to the hospital but had to put on scrubs and assist Serge in a lung-transplant operation because all the other surgical staff is on a field trip to the local water park. This scene, to the best of my knowledge, is fictitious.

> SERGE: OK, now clamp it.
> PETE: Are you telling me to shut up?
> SERGE: No, just hand me the clamp.
> PETE: I don't see a clamp here.
> SERGE: It looks like a pair of scissors.
> PETE: Oh got it. [Hands Serge the clamp.]
> SERGE: Thanks.
> PETE: Is it weird that this is making me hungry? I've got two extra-large meat-lovers combos in my van if you want.

Simply knowing more about something can lead you to use words that others who are less expert just can't understand or at least don't understand in the same way that you do. But that's just our intuition. Compelling though the tale of Serge the surgeon and Pete the pizza delivery guy might be, it doesn't tell us objectively and empirically how differences in experience affect comprehension. And, to get right down to it, it also doesn't tell us what the brain is doing differently in experts and novices that leads to eventual differences in understanding. But there are researchers working on this.

The first question we have to ask is whether people with different degrees of expertise in some domain actually understand the same language differently. In terms of embodied simulation, there's a pretty obvious possibility. Experts should be able to mentally simulate details that novices can't even imagine. What's more, experts might access these details quickly and automatically. How can we tell if they do?

One study asked this question for experts and novices in the do-main of sports.[1] And not just any sport—hockey. The research team used the sentence-image matching methodology we've seen at several points, in which people read sentences and then have to decide whether the image depicts the sentence or not, and the image can ei-ther match some fine details implied in the sentence or not (recall the orientation of the nail or the shape of the egg). The key thing that this new study did was to include both sentences about objects that every-one would have experience with (like nails or eggs) and sentences about objects that only hockey experts would be familiar with. For in-stance, the researchers included sentences about hockey equipment and pictures of the equipment in the contextually appropriate config-uration. For example, a sentence like *The referee saw the hockey hel-met on the player* probably implies that the facemask on the helmet is clipped closed. But *The referee saw the hockey helmet on the bench* probably implies that that mask is unsnapped. In case this doesn't make any sense to you (perhaps, like me, you're from California or some sim-ilar place where hockey is as foreign as curling), you'll have to take it on faith. Apparently, you wouldn't know which picture went with which sentence unless you had significant experience with hockey.

The researchers conducted this experiment with two groups of peo-ple. The first were members of the general population with no partic-ular expertise with hockey (hockey being a distant fourth in popularity among American sports, this was presumably not a hard demographic to track down). And the other group was composed of hockey experts—people who played competitive hockey. Intriguingly, there are a couple of possible outcomes. On the one hand, it could be that expertise doesn't make a difference, so that whatever the general population does (say, they're faster to respond to the matching images for the domain-general stimuli, like eggs and nails, but not for the hockey-spe-cific ones), the hockey players do the same. This is more plausible than it might seem. Just because you know lots of details doesn't mean you necessarily access them, especially when, as in this particular task, it wouldn't help you at all to do so. (Remember that people were supposed to answer "yes" regardless of whether the configuration of the object matched the sentence, so mentally simulating object details could be more of a distraction than anything else.) But, on the other hand, it could be that expertise does make a difference. Perhaps hockey experts automatically activate the appropriate configurations of hockey objects, just as they do for the domain-general ones. But hockey novices, who know little to nothing about hockey, have only had enough of the requisite experience with the domain-general sentences and pictures to be able to simulate their details, where they show a match advantage, but they show no effect for the hockey sentences.

This second possibility is in fact what the results showed. Everyone was faster to respond affirmatively to the matching pictures for the nonhockey stimuli. But only the hockey experts responded faster to the matching hockey stimuli. Strangely, however, although they were faster to respond to the matching hockey pictures, the hockey experts were no more accurate than novices in responding. So what this means is that both hockey experts and novices understood the basic gist of the sentences, even the ones about hockey. But they were understanding them differently. The hockey players were constructing more visual detail, based on their deeper experience with the domain. This interestingly didn't lead them to do any better in the task, but it could make a difference for other tasks, where those perceptual details make a difference.

And just in case you thought that this expertise effect is specific to hockey—maybe there's a special neurological condition induced by habitual Zamboni exposure—the researchers replicated the study using stimuli about football and either members of the general population with little football experience or competitive football players. And, sure enough, they got the same effect. Expertise affects the detail with which people understand language.

Of course, this work still leaves a lot more to be done. For example, it's not clear whether the differences are merely due to differences in perceptual specificity or whether experts are also more likely to project themselves into the minds and bodies of protagonists that they identify with. (If I'm a hockey player myself, I might be more likely to simulate myself as the hockey player in the sentence.) It's also not clear whether the differences between hockey experts and novices (or football players and novices) are due to differences in individual experience or other individual differences, like maybe their degree of athleticism or their interest in sentences about sports in general. One way to tease these apart would be to do something clever like give football players hockey sentences, and vice versa. But that study hasn't been conducted.

However, one issue that has been pursued is the question of the neural basis for individual differences in comprehension.[2] What's going on in hockey players' brains while they're understanding language about hockey? How is this different from what hockey novices' brains do when confronted with the same language? And what do these brain differences translate into, in terms of differences in what hockey experts versus hockey novices are actually understanding?

A group of researchers at the University of Chicago addressed these questions in the following way. They reasoned that if experts were mentally simulating differently from novices, then the place in the brain where that would be most visible would be in the motor cortex. We know that people generally use their motor systems to construct simulations of what it would be like to perform described actions. So experts might understand language about actions that they're experts in differently from the rest of us because of increased motor simulation. Maybe, as a hockey player, you're more likely to simulate hockey actions than we hockey novices are. This difference in motor simulation should show up in the motor cortex—in the form of greater activation for hockey experts to language about hockey actions than for hockey novices.

So the researchers presented people with sentences about hockey actions, like *The hockey player finished the shot*, and non–hockey actions, like *The individual pushed the cart* while they were in an fMRI scanner. Some of their participants were hockey experts and others were novices. So, did the hockey sentences induce different patterns of brain activation in the hockey experts than in novices? People with more experience (the hockey players) displayed more activation in their left premotor cortex. As you might recall, the left cerebral hemisphere controls the actions of the right side of the body (which was the participants' dominant side, because they were all right-handed). And as we saw earlier, the premotor cortex is responsible for complex, well-learned actions and is often active during motor simulation. So hockey players showed more activation in the part of their brain that controls well-learned actions on their dominant side. This isn't conclusive evidence that this increased activation indicates increased motor simulation, but it's pretty good circumstantial evidence.

There was actually a second part to the results that was a lot more surprising. If experts were just doing more motor simulation overall, you might think that they would be using not only the premotor cortex more, but also the primary motor cortex (the part of the brain that actually sends signals to the skeletal muscles). This, however, is the opposite of what the fMRI data showed. When listening to hockey sentences, the hockey experts actually showed less activation of the primary sensory-motor cortex, on both sides of their brains, than the novices. There could be several reasons for this. Perhaps the most likely is that novices have a harder time understanding hockey sentences, so they do more detailed, low-level motor simulations of what the actions would be like to perform, whereas experts who already have lots of experience with those scenarios only activate the higher-level motor routines they've encoded in premotor cortex. The upshot, though, is that the brains of experts and novices respond differently to language about a particular activity: experts activate brain areas responsible for controlling well-learned actions more, while novices activate primary sensory and motor areas more.

But do these differences in brain activity actually mean anything for the differences in comprehension that hockey players and hockey novices display? In other words, do the differences in the activation of premotor and primary motor areas actually explain the fact that hockey players activate detailed embodied simulations of hockey

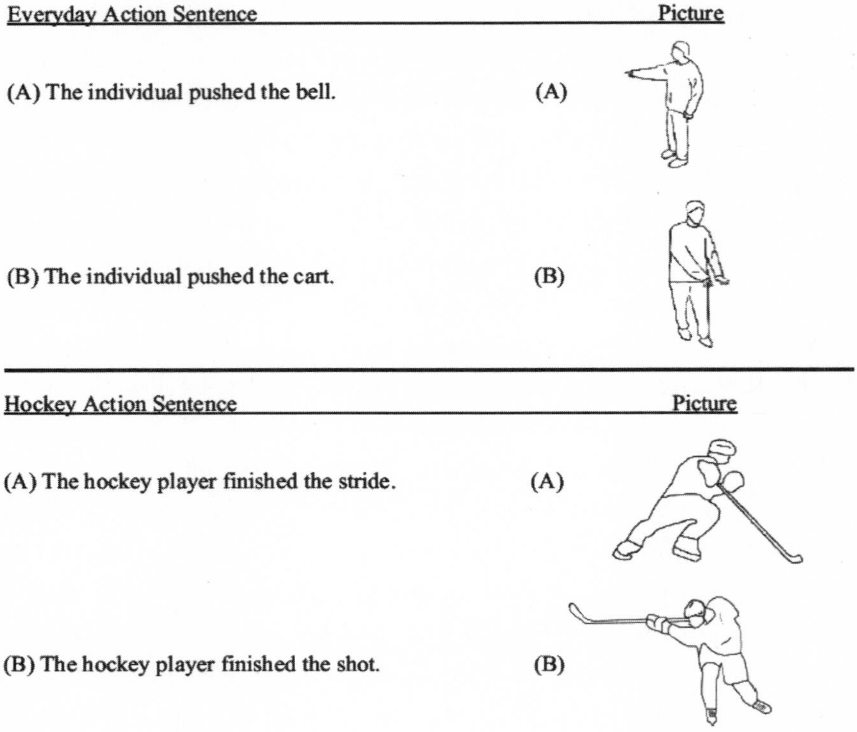

Everyday Action Sentence   Picture

(A) The individual pushed the bell.   (A)

(B) The individual pushed the cart.   (B)

Hockey Action Sentence   Picture

(A) The hockey player finished the stride.   (A)

(B) The hockey player finished the shot.   (B)

FIGURE 41

scenes, while hockey novices do not? One way the same research team looked at this was with a second task, using the same participants. After listening to all the sentences in the fMRI, the participants performed a sentence-image matching task, on both hockey and non-hockey sentences, as you can see in Figure 41.

Their logic was to use statistical tools to see, across all the participants in the study, how well experience predicts brain activity and how well brain activity predicts behavior on the sentence-image matching task. So they looked to see how well hockey experience predicts activation of motor brain areas (pretty well). Then, because these same participants performed the sentence-picture matching task after their fMRI scans, they looked to see how well activation of these brain areas in the scanner predicts each participant's match advantage in the sentence-picture matching task (pretty well). Then they looked to see whether, taking these factors into account, hockey experience carries any additional predictive value on match advantage over brain activation (it doesn't). From this, we can conclude that domain-specific experience works through differences in brain activation of motor areas to produce the faster responses to matching than mismatching pictures depicting hockey scenes. In other words, the qualitative difference in understanding language about actions by experts versus novices is at least in part mediated by differences in how these people use their motor systems during comprehension.

These results are pretty compelling, but they do leave one lingering concern. Hockey players differ from novices in a lot of ways. One is, by definition, their degree of experience with hockey. But it's also possible that they differ in other ways, for instance, their degree of athleticism, interest in sports in general, aggressiveness, or how many of their original teeth they still have. Some of these traits might be more relevant to language understanding than others, but the larger point is that because hockey players and hockey novices in the studies described here are drawn from different populations, there's no way to ensure that it's their hockey experience itself, and not other factors, that's responsible for the difference in their behavior and brain activity.[3]

But you can get around this problem. To ensure that it's experience itself that makes the difference, you'd need to take people from a single population, say, undergraduate psychology students, and then put them randomly in two groups, one that gets one kind of experience, and another that gets a different type of experience. If these two populations understand language differently, then you can be pretty certain that it's experience, and not other individual differences, that's responsible.

That's the tack that a recent study took.[4] People came into the lab and performed two tasks. The first involved matching words to pictures of objects. The pictures depicted the objects oriented in different ways; half of the participants saw a given object, like a toothbrush, say, aligned horizontally, and half saw it depicted vertically. This was the experience manipulation. Then they performed an irrelevant task to distract them; it was a mental rotation task like the one we saw in Chapter 3. And then, finally, they read sentences, presented one word at a time, while their eyes were being tracked. The trick was that some of these sentences mentioned the objects they had previously seen pictures of. And they implied that these objects were in orientations that either matched or mismatched their orientations in the earlier pictures. For instance, *Aunt Karin finally found the toothbrush in the sink* might imply a horizontal orientation, but *Aunt Karin finally found the toothbrush in the cup* implies a vertical orientation. The experimenters hypothesized that people would have a slightly harder time reading sentences that implied an orientation for an object that was different from the one they had seen twenty minutes before, even though the experiments were supposedly unrelated.

In Figure 42, you can see their eye-tracking data. Mostly people took the same amount of time to read the different parts of each sentence. But the big difference pops up in the fourth segment, where the critical object's orientation is implied. People spent about 50 milliseconds longer reading this part of the sentence when it mismatched the previously seen picture than when it matched. Longer reading times like this indicate that people were having greater difficulty with the segment in question, so this result seems to indicate that experience matters to comprehension. Even seeing an image of an object just once has an effect on how you understand subsequent language. Probably, during the sentence reading task, people were reactivating visual representations of the objects they had previously seen, and, when these conflicted with the perceptual characteristics, namely the orientation, of the object they were now reading about, it took them longer to integrate the new information they were reading into their ongoing representation of what the sentence was about.

Although the exposure that people in this training experiment got was much more limited than the kind of experience that people have in developing hockey expertise, still it had an effect on comprehension. So if even this limited degree of experience can affect embodied

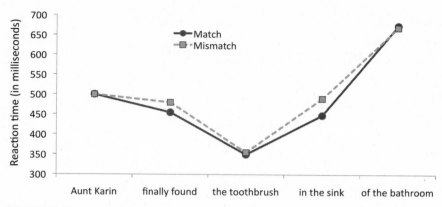

FIGURE 42

simulation processes in understanding across individuals, it's certainly reasonable to believe that extensive training over the course of a lifetime, say, in hockey, for instance, can pack even more of a punch.

The role of experience in shaping simulation and comprehension is an important discovery. We're all experts in certain things but not others, and, as a result, our embodied simulations across these different domains are of variable specificity, detail, and accuracy. Precisely which domains are which differs from person to person, which makes communication something of a minefield. From an even broader perspective, expertise is just one way in which the unique experiences we have contribute to the shape of our cognitive systems and the way that we understand language. It wouldn't be surprising if differences in the experiences we've had other than expertise—individual but formative moments like being bitten by a dog once as a child, for example—had just as big an impact on the meanings our minds make.

Different Strokes

People differ not merely in the particular experiences they've had but also in what they're capable of. Differential psychologists (people who study differences across people) break it down like this. People differ first in what they are able to do cognitively—their *cognitive abilities*.

Some people are very adept at doing mental rotation in their heads. Others can remember whole sonatas from memory. But there's a second, more subtle difference among people, and it is at least in part built off of their different cognitive abilities. People also simply have different preferences for how to use their minds, for instance, in how they perform certain cognitive tasks. This is known as your *cognitive style*. We're first going to look at some key ways in which people vary in their individual cognitive styles. Then we'll see how people with different cognitive styles understand language differently.

The most relevant dimension of cognitive style for our purposes has to do with the extent to which people use their vision systems to perform tasks other than vision. Although we've sort of been taking it for granted so far that everyone performs visual simulation, there's actually quite a lot of variability across individuals in the extent to which people do so. At the most extreme end, about 3 percent of people report never having visual imagery at all.[5] And this is among people with normal vision. Whether or not their intuition about their conscious experience of imagery is also indicative of how they unconsciously use embodied simulation, there's lots of clear evidence that there are people who tend to perform cognitive tasks through language rather than through simulated vision. These people are often called *verbalizers*, contrasted with *visualizers*.[6] Here's what this distinction means. Suppose you're asked to memorize the string of beads in Figure 43. Go ahead. There will be a test later.

FIGURE 43

Got it committed to memory? If so, reflect for a moment on how you tried to deal with this memorization task. Some people—visualizers—commit the image to visual memory so that they can recall it later through simulation. Others—verbalizers—will memorize a sequence of words characterizing the shade of the beads, which they can retrieve later.

Of course, most people are well equipped to either verbalize or visualize; the distinction between these two groups appears to be more a difference in cognitive style than ability per se. There are plenty of tasks on which visualization will be more effective than verbalization, or vice versa. For instance, in a property verification task (does a horse have a mane?), the exact relation between the object and the property influences how likely one is to use verbalization versus visualization. If the two words are strongly associated—that is, if you say or hear them together a lot—then you're more likely to just rely on the words.[7] For instance, to determine if *mane* is a property of *horse*, you probably don't need to do much visualization. However, for a less frequently pair of words, like *mane* and *pony*, you might be more likely to visualize. This doesn't result from any intrinsic difference between horses and ponies—it's just that in one case word associations do the job, and, in the other, imagery is more useful.

There are even ways that the two processes—verbalizing and visualizing—appear to affect each other. For instance, if you have people memorize pictures and then later draw them from memory, the word that appears with the image affects the character of the subsequent redrawing, as shown schematically in Figure 44.[8] If the top

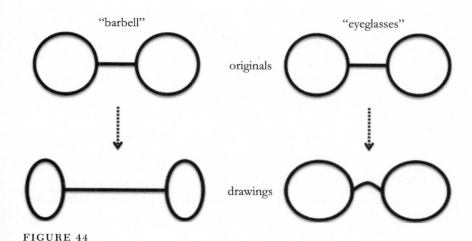

FIGURE 44

drawing, of two circles linked by a line, is labeled *barbell*, people later redraw it from memory differently than when the original is labeled *eyeglasses*. The point is that your embodied simulation of something you perceived isn't independent of the words you use to verbalize what that image represents, and both processes may be active in memory, just to varying degrees.

So we have visualizers who tend to visualize—that's a process that all the work on visual simulation positions us to understand pretty clearly. But what is this verbalization thing? Verbalization is the use of language to perform cognitive tasks. It can be either overt or covert. *Overt verbalization* is actually saying words out loud or writing them down as part of your solution. So you might have overtly verbalized when memorizing the beads if you actually said words describing the shades of the beads in order. *Covert verbalization* would have been just to say (and hear) them silently in your mind. Overt verbalization uses your motor and auditory systems to create perceptual and motor memories of the sound and feel of the words you're articulating. Covert verbalization also uses your motor and auditory systems, just to simulate what it would be like to say and hear the relevant words.

How do we know that covert verbalization uses the motor and auditory systems? From some ingenious studies.[9] Imagine that you had a device that could selectively disrupt the activity of a specific part of the awake brain. No, I don't mean a pick axe. I mean a harmless, non-intrusive tool that would scramble the neurons in a tightly circumscribed bit of brain very briefly. Such a tool exists: Trans-cranial Magnetic Stimulation, or TMS. The way TMS works is by taking a powerful electromagnet, applying it to a specific spot on the scalp, and allowing it to induce a strong and targeted but temporary and harmless magnetic field that transiently interferes with normal neuron functioning at a specific spot in the brain. Perfect. Now what do you think happens when you apply TMS to the brain areas responsible for moving the speech articulators—the mouth, tongue, and so on? Well, one obvious result is that people have trouble talking. After all, you're knocking out the neurons that control those muscles. But the remarkable thing is that TMS applied to the speech centers also interferes with your ability to covertly verbalize words.[10] This means that inner speech, despite being totally silent and unarticulated, uses the parts of motor cortex that control your mouth and tongue. When you covertly speak to yourself—even though you're not actually moving your mouth—you activate brain regions that are responsible for controlling your speech articulators.

One of the more impressive applications of this fact has been the development of computer interfaces that can basically read your mind by minding your body. This technology was developed by NASA scientists who were interested in how to allow people to control a computer in situations where they can't type or perhaps even speak.[11] You can imagine, for instance, that, aboard an accelerating rocket, the astronomical amount of force on your body might make it impossible for you to move much of anything, even your mouth. Fortunately, as we now know, when you covertly pronounce words, there is activation in your motor cortex—not as much as when you actually move your mouth, but enough to send a measurable electrical signal to the muscles of your face. And as a result, different words produce different, although very subtly different, patterns of electrical activity in the face muscles. So the NASA researchers placed sensitive electrodes on people's throats and had them to just covertly verbalize different words, like "right," "left," and so on. The researchers then trained a computer to recognize the differences in motor patterns that reflected the different words. Ultimately, the software was able to discriminate with 92 percent accuracy among six different commands, which under the right circumstances you could imagine mapping onto commands that NASA might find important, like "stop," "go," "right," "left," "fire photon torpedoes," "activate flux capacitors," and so on. Now, you might think that 92 percent is good but not quite up to par for something as important as navigating a space vessel, and, for that reason, research continues (*I'm sorry, Dave, I thought you told me to jump to light speed*). But the take-home message is that even when you're verbalizing covertly, you're using your motor and perceptual systems. So the difference between visualizers and verbalizers isn't that visualizers mentally simulate while verbalizers don't. It's that they simulate different things (visual percepts versus words) in different modalities (visual versus motor and auditory).

And if the distinction between verbalizers and visualizers weren't enough, there are even finer distinctions among different sorts of visualizers. Spatial visualizers tend to reason about the arrangement of objects in space and their motion (think about the function of the Where pathway), while object visualizers are good at perceiving and recalling visual properties of objects themselves (as though they were relying more heavily on the What pathway). There is a small but consistent tendency for men to be stronger spatial visualizers and women to be stronger object visualizers, though this is far from universal.[12] And it gets even more complicated, because people's spatial and object visualization preferences are uncorrelated. So you could be strong at either one type of visualization or both or neither.

Cognitive style is a good predictor of your performance on particular tasks. Object visualizers outperform others on tasks specifically related to perceiving and recalling objects. For instance, there's an outline drawing of an everyday object in the degraded picture in Figure 45.[13] Time yourself while you try to find it. Object visualizers see the object about 70 percent of the time, as compared with others, who only see it closer to 50 percent of the time. And object visualizers are also faster, at an average of about 8 seconds, as compared with closer to 12 seconds for others. (In case you're stumped, there's a hint that might help you here.[14])

We're now in a position to come back to the question of whether people with these different cognitive styles tend to understand language differently.

Suppose two people—one who does relatively little object visualization but lots of verbalization and another who does the reverse—read the following passage (which you might recognize, because it's from *Jane Eyre*, by Charlotte Brontë):

FIGURE 45

The red-room was a square chamber, very seldom slept in, I might say never, indeed, unless when a chance influx of visitors at Gateshead Hall rendered it necessary to turn to account all the accommodation it contained: yet it was one of the largest and stateliest chambers in the mansion. A bed supported on massive pillars of mahogany, hung with curtains of deep red damask, stood out like a tabernacle in the centre; the two large windows, with their blinds always drawn down, were half shrouded in festoons and falls of similar drapery; the carpet was red; the table at the foot of the bed was covered with a crimson cloth; the walls were a soft fawn colour with a blush of pink in it; the wardrobe, the toilet-table, the chairs were of darkly polished old mahogany. Out of these deep surrounding shades rose high, and glared white, the piled-up mattresses and pillows of the bed, spread with a snowy Marseilles counterpane. Scarcely less prominent was an ample cushioned easy-chair near the head of the bed, also white, with a footstool before it; and looking, as I thought, like a pale throne.

To an object visualizer, rich description like this could be a virtual feast for the senses. A verbalizer, however, might find less in it to be enraptured by and might end up skipping down the page to find the next bit of dialogue. That's what I did.[15]

By contrast, verbally clever banter—without any real visual detail—might be more likely to attract and hold the attention of a verbalizer (but not a visualizer). For instance, consider the following exchange from *The Importance of Being Ernest*, by Oscar Wilde, in which Algernon is speaking with his servant Lane:

ALGERNON: [. . .] Oh! . . . by the way, Lane, I see from your book that on Thursday night, when Lord Shoreman and Mr. Worthing were dining with me, eight bottles of champagne are entered as having been consumed.

LANE: Yes, sir; eight bottles and a pint.

ALGERNON: Why is it that at a bachelor's establishment the servants invariably drink the champagne? I ask merely for information.

LANE: I attribute it to the superior quality of the wine, sir. I have often observed that in married households the champagne is rarely of a first-rate brand.

ALGERNON: Good heavens! Is marriage so demoralising as that?

LANE: I believe it *is* a very pleasant state, sir. I have had very little experience of it myself up to the present. I have only been married once. That was in consequence of a misunderstanding between myself and a young person.

ALGERNON: [Languidly.] I don't know that I am much interested in your family life, Lane.

LANE: No, sir; it is not a very interesting subject. I never think of it myself.

ALGERNON: Very natural, I am sure. That will do, Lane, thank you.

LANE: Thank you, sir. [LANE goes out.]

ALGERNON: Lane's views on marriage seem somewhat lax. Really, if the lower orders don't set us a good example, what on earth is the use of them? They seem, as a class, to have absolutely no sense of moral responsibility.

Did you nod off during this passage? If so, you may well be a visualizer, because there's really nothing here to keep your attention. For the verbalizer, though, there's a smorgasbord of novel and counterintuitive linguistic tidbits to gobble up.

In principle, it could be that people with different cognitive styles not only harbor different preferences for vivid visual description versus witty banter, but that, when forced to read a given text, they also display different capacities for encoding and recalling its content, depending on what it's about. So you might expect that if you asked a bunch of visualizers to read the passage from Jane Eyre, they would construct more vivid visual simulations of the red room than verbalizers would. (Did you not even know that the room was red? Verbalizer!) And later on, if you asked them comprehension questions, they might do a much better job at correctly reporting what objects in the room were located where, and what color they were. By contrast, you'd expect that verbalizers should do a better job at both tracking and remembering who made exactly which witty crack to whom in the Oscar Wilde scene.

There's very little empirical work on how differences in cognitive style affect comprehension. But one recent study stands out in that it relates properties of people's embodied simulations during intentional imagery and during language comprehension.[16] There's an interesting visual illusion called the motion aftereffect. You might have experienced it before. It happens to me when I'm hiking in the woods for a while. When I come to stop, the trees look for a brief instant like they're moving forward, away from me. People also report that they have the effect when staring at a waterfall for a while. When they then turn to look at something stationary, say, the mossy wall next to the waterfall, it appears to be moving upward. In both cases, and in general, the motion aftereffect is the visual experience of things moving in the opposite direction as something you've just adapted to.

The explanation for why adaptation like this occurs is complicated, but the basic idea is that we have neurons that detect motion in various directions, housed in the dorsal visual stream—the Where pathway. When we're not experiencing motion in any direction, these neurons are competing with each other, but none is more active than any others. However, when we perceive motion in a particular direction, the appropriate neurons fire. But if they're continuously stimulated for an extended period of time, they start to fire less. This is the adaptation part. The motion aftereffect comes into play when we quickly stop looking at something moving and turn to something stationary. Our neurons that detect motion in that same particular direction are now not firing because they've adapted to motion, so the neurons that detect motion in the opposite direction can take over. They're going to be somewhat more active than these adapted neurons, so we have the experience of motion going in the opposite direction.

Now, it's known that we can experience a motion aftereffect not only when we watch motion for a while but also when we imagine motion for a while.[17] This shouldn't be too surprising—we already know that intentional visual imagery uses the vision system. But the key finding of interest with respect to individual differences is that people differ substantially in the extent to which mental imagery produces a motion aftereffect. Some people display effectively no motion aftereffect, while, for others, it's quite pronounced. If the magnitude of the effect is an indication of the extent, strength, or specificity of people's visual spatial imagery, then it serves as a suitable measure of the extent to which they are strong visualizers of motion (a component of what goes into making someone a spatial visualizer). Do stronger versus weaker spatial visualizers understand language about motion differently?

Researchers at Stanford figured out a way to test this. They had people listen to stories about things moving up or down. For instance, one of their upward stories was as follows:

You are running a psychology experiment in which you have trained hundreds of squirrels to race each other up a wall for a piece of food. Now you want to see what happens when they are all released at the foot of the wall at once. You watch through a small window in the next room as the cages are opened and the squirrels leap onto the wall in a frenzy. The little fur balls scurry up the wall in one relentless stream, despite obvious defeat in the race. Zip! The brown creatures surge up the wall with amazing agility. You see the same behavior in squirrel after squirrel—one swift jump onto the wall and an instantaneous burst upward. Zoom! The squirrels rush up the wall like a giant current. As if in a trance, the squirrels swiftly stream past your eyes in their race for the top of the wall. Zoom! More and more squirrels jump onto the wall and scurry upward. You watch them course up the wall in a blur. The squirrels continue to sprint upward in a flash. They spout onto the wall and surge directly toward the top. Your eyes remain focused on the mob of squirrels teeming up the wall. You can no longer pick out individuals as they dash for the top.

The downward stories were similar, but of course described downward motion.

To measure the extent of the motion aftereffect that these passages induced across people, each narrative was interrupted at intervals so that the participants could perform a secondary task. This was a moving dot task. The basic idea of this methodology is to have people look at a screen showing a lot of dots, a hundred of them. The dots are moving in various directions. When the dots are moving randomly, you don't see them as moving coherently in any particular direction. But when enough of them are going in the same direction—and the coherence threshold differs across people—you can see them tending to move in that direction. The motion aftereffect is known to interfere with your ability to perceive coherent dot motion. If you've adapted to upward motion, you're more likely to say that the dots are moving downward, and vice versa. So when people are experiencing a strong motion aftereffect from reading the passage, they should show a large interference on being able to perceive the moving dots.

Using the moving dots paradigm as a measure of motion adaptation, the researchers had people perform two tasks. The first measured the magnitude of their motion aftereffect during visual imagery—people had to imagine motion up or down and then saw moving dots and had to say which direction they were going. The second task was the same, except that, instead of imagining motion, they read the passages about upward or downward motion and then performed the dot task. The idea was to see whether people who show stronger motion aftereffects from imagery also show stronger motion aftereffects from language comprehension. On the one hand, it might be that nobody shows motion aftereffects from language comprehension. For a motion aftereffect to occur, people would have to be systematically using the particular neurons in their dorsal Where pathway that detect motion in a particular direction to the point where those neurons start slowing down. On the other hand, it could be that everyone shows a motion aftereffect. Perhaps individual differences in cognitive style don't matter much at all to language comprehension. Finally, on the third hand, it could be that people who have a stronger motion aftereffect with imagery also have a stronger motion aftereffect during language comprehension.

And that final possibility is actually how the data turned out. Not everyone showed a motion aftereffect from reading about the teeming squirrels and so on. But some did. And there was a positive correlation across the participants between their degrees of motion aftereffect produced by imagery and by language. The more their imagery used the Where pathway, the more language about motion did, too. So individual differences in how people use their vision systems for things other than language, like imagery, make a difference for how people also use their vision systems in language. Spatial visualizers just might create more vivid or longer-lasting embodied simulations of motion when they're reading language about things moving than people with other cognitive styles.

Of course, there are other ways that people with different visualizing styles engaging different cognitive processes could plausibly understand the same texts in a way that makes a difference for what they actually get out of them. Suppose you take two people—one an object visualizer and the other a spatial visualizer. And you have them read a text that describes a physical system, like a mechanical clock or an ecosystem. The spatial visualizer might do a much better job of picking out the spatial relations among objects and how they move relative to one another. The object visualizer, though, might have more success tracking the visual properties of the various objects. So when you later ask them what they understood, they might be able to access very different aspects of the text. We don't actually know whether this is true yet. It's an experiment waiting to happen.

Failures to Communicate

The empirical evidence we have now only scratches the surface; there could be even more dramatic differences in meaning across individuals than we've seen. The question at hand is what the potential ramifications are if people do in fact understand language differently. Could differences among individuals affect our ability to understand each other? After all, when we use language to communicate, we are using a channel of quite limited bandwidth—just a few words in a particular order—to convey deeply complex and richly detailed experiences. To the extent that the way we go from these experiences to words as speakers or from words to experiences as listeners is different from one person to the next, communication might be more likely to break down. So we might wonder whether having a different cognitive style or a different degree of expertise with a domain lead to less efficient, less successful, or less enjoyable conversations between people. Does communicating across differences in experience and cognitive style cause more failures to communicate?

It's a fascinating idea, but, simply stated, we don't know. Fortunately, it wouldn't be hard to test. Suppose you have people come into the lab two at a time. The people don't know each other, and they're randomly paired together. Because people vary naturally in cognitive style and in expertise, you'll sometimes end up with people who are well matched along these dimensions and other times with people who aren't. What if you had these pairs perform some collaborative task? Say, one person sees a set of blocks arranged in a particular configuration and has to tell the second person how to assemble a different set of blocks into the same configuration. The blocks are all visually unique; there are long ones, short ones, ones of different colors, and so forth. But they also all have distinct words written on them. So the players can complete the task by referring to the distinct blocks using visual descriptions, or they can use the words written on them.

Some of the pairs will settle on one strategy, while others will settle on the other. But if communication improves when people have similar cognitive styles, we might expect to see a couple things. First, pairs who share a cognitive style should settle on a solution that exploits their shared preferences. Second, it should take heterogeneous-style pairs longer to settle on a shared strategy than homogeneous-style pairs. And, finally, it should be harder for heterogeneous-style pairs (measured in terms of how long the task takes and how accurately the configuration is replicated) than for homogenous-style pairs. It might also be less enjoyable for heterogeneous-style pairs. This is all hypothetical of course, but it needn't be. This is a testable idea and one that would reveal the extent to which individual differences in how we make meaning affect our ability to communicate with each other.

Finally, because I promised, here's your test. Without flipping back to the original image, is this the string of beads you memorized before?

FIGURE 46

How do you know? Did you verbalize a list of names, or visualize the locations of the shades? As you now know, in one respect it doesn't matter whether you verbalized or visualized—either way, you used simulation. But in other respects, like in terms of the mechanisms you used or your ability to communicate with others, it might matter a lot.

CHAPTER 8

Lost in Translation

It was a relaxing vacation touring the English countryside, but, as all good things do, it has now come to an end, which means that you have to return your rental car. You park it and step into the rental agency office. After a minute or two, one of the customer service representatives comes to the desk and says that there seems to be damage to your car—a large dent on the driver's side. Perturbed (What dent? You didn't hit anything, did you?), you look out the window at the car, which looks something like what you see in Figure 47.

FIGURE 47

Just like you thought. No damage. What gives?

It's possible that the rental place made a mistake. Maybe they mixed up which car was which? That could explain it.

But suppose you compare notes with the agent, and, sure enough, the car you returned, the very same car that you're looking at, is the one the agent says is damaged. So now you're in a face-off. You insist the car isn't damaged—just look at it! The agent is completely certain that there's damage on the driver's side. One of you is wrong. It's got to be him, right?

Right. Except for one little thing. All of a sudden, you remember. You're in England. And that changes everything.

That's because, in England, and some other places like Japan and Australia, you drive on the left-hand side of the road, and the driver sits on the right side of the car. The driver's side in England is actually the right side of the car, which you, unfortunately, can't see in the image on the previous page. There could well be a dent there, hidden from view. Ahem.

Maybe this is what George Bernard Shaw meant when he said that England and the United States are two countries separated by a common tongue. On the surface, it seems that we should be able to communicate with one another just fine. Of course, there are the well-publicized exceptions. A *boot* is part of a car in England and they wear *pants* under their *trousers*. There are even some less-publicized differences. Don't, for instance, make the mistake I did and toss the verb *toss* around in public. (It means . . . well, look it up.) But all in all, we and our British counterparts mostly use the same words, and we have very similar grammar. And yet when we talk about the *driver's side*, something's amiss. At some level, the *driver's side* is the very same thing here and there. It's the side where the driver sits. But, at a deeper level, how we flesh that meaning out is different—we have different visual representations of what the *driver's side* looks like and different expectations about where to find it. This leads us to act as though we don't understand each other at all, in large part because our assumptions about how communication normally works break down.

What's going on here is a larger-scale version of the effects of individual experience that we looked at in the last chapter. We interpret language in terms of the experiences we've had. And in different cultures, people have different experiences. This is part of what we mean by culture: similar experiences among people in a particular group that might differ from the experiences of members of another group. Even though *driver* and *side* mean very similar things for British English and American English speakers, we interpret those words, constructing embodied simulations, based on our culturally contingent experiences. When those experiences are different enough, it occasionally makes it quite hard and, in some cases impossible, for us to understand one another.

It Means Squat

As we saw in the last chapter, the embodied simulations we construct when understanding language depend on the experiences that we've personally had. When those experiences differ systematically across cultures, this can in principle lead to the same words being interpreted differently—the same words can drive different embodied simulations for different populations of people.

For example, take the sentence *All afternoon, I was waiting for my brother on the corner.* In understanding a sentence like this, you might simulate what waiting on a corner would look like. Or if you identify strongly with the speaker, then maybe you simulate what it would feel like to wait. But either way, you're likely to simulate standing, pacing, or sitting, and, if you're sitting, then it's probably on a bench or maybe on the curb. Those are pretty much the options. At least, if you're an American.

The thing is that the way we stand and sit and wait aren't universal.[1] There are plenty of other ways to hold your body, and there are cultures where standing and sitting aren't the default waiting positions. For instance, in China, when people don't want to stand, they often squat. They squat on the ground, they squat on the sidewalk, and they

FIGURE 48

sometimes even squat on benches, rather than sitting on them. To the typical American, this is surprising and can even look uncomfortable. But it's just what Chinese people do. The point is that people use their bodies differently to wait in different cultures. And, as a result, understanding language about waiting could involve very different embodied simulations if you're culturally Chinese or culturally American.

The same is true of lots of other types of action. People eat differently around the world (with their hands, with chopsticks, with cutlery, etc.). They pray differently (Hands together? Knees on the ground? Sign of the cross? Rocking back and forth? Speaking in tongues?). They even—how to put this delicately—execute their bathroom functions in different positions and clean up differently afterward. The result is that the very same words (*eat*, *pray*, *poop*, and so on) will mean radically different things depending on your cultural experiences.

Here's another East-meets-West example. If you're an American and you want to hand someone your business card, you'll likely pick it up with one hand (probably your dominant one) and extend it to the intended recipient. But cultures have specific prescriptions for how to use your hands to give and receive small objects. In Japan and Korea, for instance, the status of the person you're interacting with determines how many hands you use. For someone of lower status than

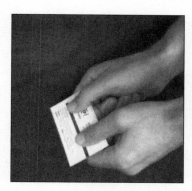

FIGURE 49
The polite, two-handed
way to give small objects
to people in Korea (as well
as Japan and China).

you, say a child or a subordinate at work, you can use one hand, just like an American would. But, for people of higher status, like your boss or a visiting dignitary, you'd have to use both hands. Using just one hand with a high-status person would be a blatant violation of politeness norms.

The different conventions that Americans and Koreans have for giving objects to people, just like the different ways people wait across the world, could have consequences for meaning. Americans and Koreans might understand language about giving and receiving objects using different embodied simulations because of their different physical experiences giving and receiving objects. We actually have some evidence that this is the case, from some research in our lab.[2] We had people listen to sentences that described giving actions, like giving someone a business card or an apple or a cup of tea. We manipulated whether the recipient was a high-status person (like a CEO) or a low-status person (like a young nephew). What people in the experiment had to do was to decide whether the sentence made sense or not (half of the time, we gave them nonsense sentences). The trick, though, was that they had to indicate their judgments about the sentences using one hand or two. You can see how we did this in Figure 50.

We had people sit in front of two keyboards, arranged as you see them in the figure, so that the two round yellow buttons were right in front of them. They had to press these two buttons to hear a sentence, and when they were ready to decide if the sentence made sense or not,

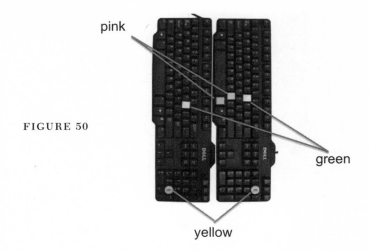

pink

FIGURE 50

green

yellow

they had to press either the pink buttons, with just their right hand, or the green buttons, with both hands. We manipulated whether the green buttons or the pink buttons indicated that the sentence made sense. As a result, people had to press either one button or two buttons to indicate that a sentence made sense. And, because the sentences described giving things to people of high status or low status, we were able to measure whether people, Koreans and Americans, responded faster with one hand or two to sentences about giving things to high- or low-status recipients. The basic idea was that the status of the recipient should matter to Koreans; perhaps they would press two buttons faster after a sentence with a high-status recipient, and one button faster after sentences about giving things to low-status recipients. But Americans, we hypothesized, shouldn't show any such effect. (If people's embodied simulations of actions depend on their cultural experiences, that is.)

What we found was a mix of the expected and the not entirely expected. For English speakers, as predicted, there was no difference in reactions times when people were listening to sentences about high- or low-status recipients, regardless of the number of hands they were using to respond. This makes sense, because Americans typically use one hand to transfer small objects, regardless of status. By contrast, as you can see in Figure 51, the Korean speakers were affected.

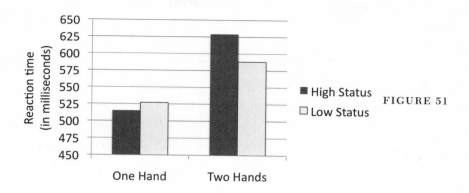

FIGURE 51

When they were responding that sentences were meaningful with two hands, Koreans were slower when the sentence was about giving something to someone of high status than low status. And the pattern was reversed (though weaker) for one-hand responses. So what we found was that the status of the recipient does indeed matter to Korean speakers. And it appears to affect their embodied simulations of action.

The odd thing is that the effect is backward from what we originally expected. Why would Koreans take <u>longer</u> to press buttons with two hands after a sentence about an action that would require two hands to perform? Our current best guess is that it has something to do with timing; in Korean, the verb comes at the end of the sentence, unlike English, and timing, as we've seen, can make all the difference in the direction of an effect like this.

This is just one example of how people understand language differently, depending on the different ways that their culture dictates they use their bodies. The embodied simulations we construct don't follow a universal template; they are deeply permeated by not only our individual experiences but the ones we gain as a result of being members of a particular culture.

The Skin of a Cat

We've just seen that people can interpret similar sequences of words in different ways, based on different cultural practices. But how deeply into the meanings of words themselves can cultural differences reach? Are there words that have fundamentally incommensurate meanings because the cultural backdrops that people use to interpret them are themselves too different?

At its heart, this is a question about whether there are equivalent words in different languages. When you translate words from one language to another, when can you expect them to have the same effect on speakers of the two languages?

Here's an example. The domestic cat, *felis catus*, has infiltrated human societies from Australia to Alaska. But the way people interact with cats is heterogeneous and culturally contingent. In some places on earth, people treat cats like quasi-human family members. Cats are fed, housed, and groomed by human servants, who even go so far as to manually clean up their waste. Cats in these feline Shangri-Las, for instance, Manhattan or Paris, largely live long, plump, healthy, carefree lives. By contrast, there are other places on earth (among them towns in the Peruvian Andes, southern China, and rural Switzerland), in which cats are used for food. They are slaughtered and their carcasses processed for human consumption in much the way that rabbits, or, more familiar to us, chickens and pigs, are elsewhere. As a result, people in these places have much broader experiences with cats: roaming free but also in processed but uncooked form as well as on their plates and in their mouths.

Now, regardless of what language the people in these cat-loving or cat-consuming places speak, their words for *cat* will be fundamentally different in what they mean, in that the same (or, roughly equivalent) words will provoke profoundly different, perhaps even incommensurate embodied simulations.

To see why, consider how different the knowledge is that you evoke when you read the word *poodle*, as contrasted with the word *lamb*. The first you probably don't eat, the second you might. How do you understand these words? Well, some of the knowledge that you access is probably quite similar; upon reading the words, you probably access visual and auditory simulations of each. *Lamb* evokes an image of a mid-sized animal with curly hair that bleats. *Poodle* evokes a visual simulation that might be quite similar to a lamb, but is certainly accompanied by different auditory simulation. Critically, though, if you eat lamb, the word *lamb* probably also leads you to access knowledge about the taste and smell of lamb meat. And you also know what lamb feels like, tactilely, in your mouth and under your knife. *Poodle*, presumably, not so much.

Assuming this characterization is accurate, the difference for you between *poodle* and *lamb* might be akin to the difference we would observe in how people from cat-loving and cat-consuming cultures understand the word *cat* or its analogues. Cat-consumers most likely construct oral-tactile and gustatory simulations when understanding what *cat* means. But people from cat-loving cultures are unlikely to ever construct such simulations in understanding the word *cat* (unless, of course it's in the context of a description of cat-eating). The difference in cultural experience can have a dramatic effect on meaning. (You will notice that I did not include pictures illustrating this particular cross-cultural difference. And you're welcome.)

In the same way that hockey experts understand language about hockey-relevant actions and objects differently from hockey novices, so it similarly stands to reason that cross-cultural differences could well produce systematic differences in meaning. As of yet, unfortunately, we don't have any empirical evidence on this issue. You could imagine an experiment looking for culture-specific incommensurability in the simulations that people from different cultures construct when processing the very same words. But for the time being, the idea that cultural practices can change the individual meaning of a word so radically is still a hypothesis.

Language as a Cultural Factor

When cultural norms and practices produce distinct enough experiences—whether it's in how people use their bodies to wait or how they interact with cats—people understand language differently. And cultures can differ in lots and lots of ways. But one critical way that we haven't considered yet is what people have to do in order to interact with the language itself. As we'll see, the way you think about events can be affected by how you interact with your language.

There's remarkable diversity in the world's languages. One of the key ways in which languages systematically differ is how they're written. Of course, most of the world's six thousand or so languages aren't written at all, so this only applies to written languages. Writing is a relatively new technology, compared to spoken language; although humans are believed to have started using spoken language at least fifty thousand years ago (and possibly much longer), the first evidence of writing comes from only about five to fifteen thousand years ago.[3] And although most people on earth now display basic literacy, even today, there remain a dozen countries (all of which turn out to be in Africa) where a majority of the population still cannot read or write.[4]

However, when languages _are_ written, they have to be written in some direction or another. And, over time, it has come to quite frequently be the case that languages are written, regardless of their script, from left to right. There are some good reasons for this. For one, when a right-handed person writes, going from left to right ensures that the ink she has just laid down or the marks she has just made in clay will not be rubbed away as successive characters are written. This ends up not working out so well for left-handers, but because they make up only about 10 percent of the population, they're forced to smudge their way from word to word or else contort their hand into an awkward and sinister crook. Long story short, most languages are written left to right on the page or the screen.

But not all languages are written this way. Some languages, like Hebrew and Arabic, are written right to left. Traditionally, Chinese was written top to bottom, and then right to left. So although there's a statistical preference for writing systems to go in one direction, people write and read in different directions, depending on the language,

English, Italian   **Arabic, Hebrew**   **Chinese (trad.)**

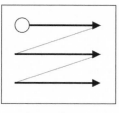 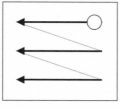 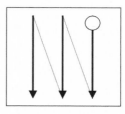

left-to-right   *right-to-left*   *top-to-bottom*

FIGURE 52

and, as far as we know, there are no global cognitive advantages to any one direction or the other.

Recently, people have started to ask whether it's possible that the direction you read and write affects how you understand language. For instance, suppose I tell you that yesterday I was in the park, and I saw a guy in a purple track suit jog past me. Which way did you see him jogging in your mind's eye? In general, there's a lot of experimental evidence showing that you're more likely to have seen him going from your left to your right than any other direction. This has led some researchers to hypothesize that people have a universal bias to simulate events as going from left to right.[5] And that idea would certainly seem to be corroborated by work on people's mental representations of events, which have shown a similar bias in English, Italian, and other European languages. But there's a problem. All these languages are written from left to right. So there's a confound here— we don't know whether English speakers and Italian speakers mentally represent events going from left to right because there's a universal bias for left-to-right motion or because learning to read and write a language from left to right leads you to think about motion this way.

The way to tease these two possible explanations apart is to ask what happens when you tell a speaker of Arabic or Hebrew (who writes from right to left) about the jogger. Will she see him go from left to right in her mind's eye, or right to left? If thinking of motion as going from left to right is a cognitive universal that's pervasive across our species, then the Arabic or Hebrew speaker should act like an English speaker. But if the direction of your writing system accounts for how you think of lateral motion, then the Arabic or Hebrew speaker should have a mental representation of lateral motion that's a mirror image of what English speakers see.

A pair of European researchers took this question on, in several ways.[6] They asked speakers of Italian (which is written from left to right) and Arabic (which, again, is written from right to left) to listen to sentences about actions, like *The girl pushes the boy*. The Italian and Arabic speakers then had to either (in a first experiment) draw the event described in the sentence or (in a second experiment) look at a picture and decide if it depicted the event in the sentence. In the first experiment, the researchers measured whether the participants were more likely to draw the subject of the sentence (the girl in the sentence above) on the right or the left; in the second experiment, they recorded how fast people responded when the subject was depicted on the right or the left.

The results from these two experiments painted a clear picture of how writing can affect meaning. In Figure 53 you see the mean reaction times in the second experiment—how long it took people to say that a picture did indeed match the preceding sentence—when the subject of the sentence was depicted on the right versus the left. What you can see is that the Italian speakers, on the left, were faster to say the picture matched when the subject of the sentence was depicted on the left, and Arabic speakers (even though they, like the Italian speakers, were performing the task in Italian!), were faster when the subject was depicted on the right. And they found a similar result when they had people draw pictures; Italians tended to draw the subject on the left, and Arabic speakers drew subjects on the right.

Taken at face value, these remarkable results seem to show that people who are used to reading and writing in a certain direction tend

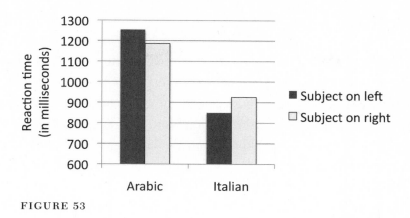

FIGURE 53

to understand language about horizontal motion as going in the same direction. But we should tread cautiously here—other cultural practices could potentially correlate with writing direction, which could muddy the explanatory waters. For instance, it could be that the visual depictions of events that Italians see in comic books, cartoons, or movies tend to cast motion from left to right, but the ones made for Arabic speakers are more likely to go from right to left. No rigorous studies that I know of have been done of the direction in which events are depicted in film across cultures. But differences in how people mentally represent events could in principle be due to how events are depicted, and not to writing direction per se. And to make things even more complicated, it's even possible that people creating comic books and other artifacts depict horizontal events in different directions in different cultures because of their language's writing direction. This would introduce another link in the causal chain. Writing direction might directly affect how people see motion in their minds, but this influence might also be mediated by cultural conventions for depicting motion.

The key thing, though, is that people who speak different languages, as part of belonging to different cultures, understand the same language about the same events differently.

Thinking for Speaking

The reason that writing direction might affect the way people think about events is because, although writing is part of language, language is itself part of culture. To learn a language is to learn a particular way of talking about the world, cutting it up into meaningful parts that we encode through the words in our particular language. The thrilling thing about looking across the world's languages is that languages cut the world up into different collections of parts. And as a result, in order to speak one language or another, you need to be making different kinds of distinctions among things in the world, cutting up the world into different categories, and paying attention to different details. To speak a language successfully, you have to think appropriately. Benjamin Lee Whorf put it more eloquently than I can:[7]

> We dissect nature along lines laid down by our native language. The categories and types that we isolate from the world of phenomena we do not find there because they stare every observer in the face; on the contrary, the world is presented in a kaleidoscope flux of impressions which has to be organized by our minds—and this means largely by the linguistic systems of our minds. We cut nature up, organize it into concepts, and ascribe significances as we do, largely because we are parties to an agreement to organize it in this way—an agreement that holds throughout our speech community and is codified in the patterns of our language. . . . All observers are not led by the same physical evidence to the same picture of the universe, unless their linguistic backgrounds are similar, or can in some way be calibrated.

Pretty words, but how much does the way your language cuts up the world really affect how you think? The skeptic in you may well think that it seems like, at the core, people are people. No matter what language they speak, people from different cultures, speaking different languages, can on average think equally complex thoughts, make equally fine perceptual distinctions, and learn new concepts and behaviors with equal capacity.

One of the most impressive differences across languages is the ways that they locate things in space. Some languages, like English, tend to prefer to use a so-called *egocentric* frame of reference. For instance, if you're taking a tour of a model home, the realtor might motion toward the bay windows and say, "And on your right is one of the most delightful features of this home." The phrase *on your right* has to be interpreted from your perspective; if you were facing another way or if you were in another place, it would refer to a totally different region in space, which is why it's called *egocentric*. However, not all languages work like this. Some languages use *geocentric* systems for locating things in space. There are different types of geocentric systems; the cardinal directions we use (north, south, east, west) are one example, but others include uphill versus downhill or uptown versus downtown.[8]

The important thing about these different spatial frames of reference is that, depending on which language you speak and which spatial frame of reference it requires you to use, you're going to have to keep track of different things. To take one example, consider the language Pormpuraaw, an Australian Aboriginal language that has recently been closely studied by psychologist Lera Boroditsky and linguist Alice Gaby. It turns out that Pormpuraaw uses a geocentric frame of reference much more pervasively than English does; so much so, in fact, that even to say "hello," you ask, "Which way are you heading?" and the response is to report your actual heading (e.g., "A short distance, north by northwest").[9]

As a result, in order to speak or understand Pormpuraaw, you need to be constantly keeping track (or at least to be able to retrieve very quickly) which direction is which. This is obviously quite different from English, where cardinal directions rarely make a difference for basic, everyday language use, like saying "hello." And so, the things you have to think in order to use English are different from the ones required for Pormpuraaw. The name for this phenomenon is "thinking for speaking." Different languages can require people to systematically engage in different patterns of thought in order to speak and understand them.

Another example of how languages make people think differently comes from color perception. Languages have different numbers of color categories, and those categories have different boundaries. For instance, in English, we make a categorical distinction between reds and pinks—we have different names for them, and we judge colors to be one or the other (we don't think of pinks as a type of red or vice versa—they're really different categories). And because our language makes this distinction, when we use English and we want to identify something by its color, we have to attend to where in the pink-red range it falls. But other languages don't make this distinction. For instance, Wobé, a language spoken in Ivory Coast, only has one color category that spans English pinks and reds. So to speak Wobé, you don't need to pay as close attention to colors in the pink-red range to identify them; all you have to do is recognize that they're in that range, retrieve the right color term, and you're set.[10]

We can see this phenomenon in reverse when we look at the blue range. For the purposes of English, light blues and dark blues are all blues; perceptibly different shades, no doubt, but all blues nonetheless. Russian, however, splits blue apart in the way that we separate red and pink. There are two distinct color categories in Russian for our blues: *goluboy* (light blues) and *siniy* (dark blues). For the purposes of English, you don't have to worry about what shade of blue something is to describe it successfully. Of course you can be more specific if you want; you can describe a shade as *powder blue* or *deep blue*, or any variety of others. But you don't have to. In Russian, however, you do. To describe the colors of Cal or UCLA, for example, there would be no way in Russian to say they're both blue; you'd have to say that Cal is *siniy* and UCLA is *goluboy*. And to say that, you'd need to pay attention to the shades of blue that each school wears. The words the language makes available mandate that you pay attention to particular perceptual details in order to speak.

The flip side of thinking for speaking is thinking for understanding. Each time someone describes something as *siniy* or *goluboy* in Russian, there's a little bit more information there than when the same things are described as *blue* in English. So if you think about it, saying that *the sky is blue* in English is actually less specific than its equivalent would be in Russian—some languages provide more information about certain things each time you read or hear about them.

The fact that different languages encode different information in everyday words could have a variety of effects on how people understand those languages. For one, when a language systematically encodes something, that might lead people to regularly encode that detail as part of their embodied simulations. Russian comprehenders might construct more detailed representations of the shades of blue things than their English-comprehending counterparts. Pormpuraawans might understand language about locations by mentally representing cardinal directions in space while their English-comprehending counterparts use ego-centered mental representations to do the same thing.

Or an alternative possibility is that people might ultimately understand language about the given domain in the same way, regardless of the language, but, in order to get there, they might have to do more mental gymnastics. To get from the word *blue* in English to the color of the sky might take longer than to go there directly from *goluboy* in Russian. Or, to take another example, to construct an egocentric idea of where the bay windows are relative to you might be easier when you hear *on your right* than *to your north*.

A third possibility, and one that has caught a lot of people's interest, is that there may be longer-term and more pervasive effects of linguistic differences on people's cognition, even outside of language. Perhaps, for instance, Pormpuraawan speakers, by dint of years and years of having to pay attention to the cardinal directions, learn to constantly monitor them, even when they're not using language; perhaps more so than English speakers. Likewise, perhaps the color categories your language provides affect not merely what you attend to and think about when using color words but also what differences you perceive among colors and how easily you distinguish between colors. This is the idea of *linguistic relativism*, that the language you speak can affect the way you think. The debate about linguistic relativism is a hot one, but the jury is still out on how and when language affects nonlinguistic thought.[11]

All of this to say that individual languages are demanding of their speakers. To speak and understand a language, you have to think, and languages, to some extent, dictate what things you ought to think, what things you ought to pay attention to, and how you should break the world up into categories. As a result, the routine patterns of thought that an English speaker engages in will differ from those of a Russian or Wobé or Pormpuraaw speaker. Native speakers of these languages are also native thinkers of these languages.

Thinking in a Language

We began this chapter by asking why communication breaks down across cultures and across languages. On the one hand, our culturally dictated experiences color how we tend to think. If we're used to holding our bodies in a particular way or seeing clothes of a particular type and so on, we will construct embodied simulations to understand language that reflect those tendencies. When two people's backgrounds are different, whether it's within or across cultures, the words they use will evoke different embodied simulations in their respective minds, and, to the extent that the task they're using language for relies on the details of those embodied simulations, then communication will be impaired.

That's because language, after all, provides a very narrow aperture through which we communicate thoughts. In speaking, we utter a few quickly selected words, seeds that are meant to produce florid gardens of meaning when they find purchase in the fertile mind of the listener. And yet minds differ because they are made over the course of individual experience. And so the soil in my mind might not make simulations blossom the same way yours does. What looks like a tomato plant in your mind might produce a seed that grows into a pumpkin when I try to interpret it.

People who speak different languages are also compelled to think differently, because of what their language forces them to attend to and the categorical distinctions it forces them make. This, of course, isn't the exclusive purview of language; people attend and perceive differently for other reasons as well. One important reason is expertise. An expert Starcraft player (Starcraft is a military science fiction real-time strategy video game that may or may not be the software equivalent of methamphetamine) sees the screen differently from a novice; he recognizes nonobvious, high-level patterns and attends to different details than the neophyte. The same is true for the expert birdwatcher, chef, or tennis player.

And if there's anything that birdwatchers do more than watch birds, that chefs do more than cook, it's use language. (It's not however a given that Starcraft experts do anything more than they play Starcraft.) To the extent that any of us are experts in anything, we are experts many times over in our native tongue. And this expertise, in the form of recurrent, obligatory attention to specific contrasts or particular features, produces the somewhat different ways of thinking characteristic of Korean, Russian, Arabic, or English speakers. We are, ultimately, what we speak.

As anyone who has ever tried to learn a second language late in life knows, it's a brutal proposition, a path strewn with the corpses of declensions mangled and tenses tortured. But, further down the path, where grammatical genders are more or less intact, where words and idioms are sturdy and sound, the more advanced second language learner often finds herself immersed in a sort of alternate universe. Yes, the words are different, but that's not the rub. More entrancingly, the farther the learner walks down into the garden of a second language, the more the world itself appears to take on different forms, not because its pieces are called by different names, but because the same tableau is seen to be composed of a different set of pieces, which crosscut and overlap the borders of the jigsaw that she is most familiar with from her native language. Part of what makes learning a second language so difficult is precisely this: the commitment one made early on in life to a particular cutting up of the world at its joints is hard to see as merely one possible commitment among many, and just as it is hard to see, it is hard to let go of.

Yet, this, ultimately, is what it is to know a language. It's to engage those habitual patterns of thought that allow you, and other people who speak the same language and have the cultural background that it references, to use it productively, naturally, fluently, meaningfully. To know a language is to mean in that language.

CHAPTER 9

Meaning in Your Grasp

For the last eight chapters, I've been clubbing you over the head with study after study, cobbling together my case that, in understanding language, we use our perceptual and motor systems to run embodied simulations. That's all fine and good, and let's suppose you've even swallowed that idea. Nevertheless, it might have occurred to you that I've been focusing mostly on language about concrete stuff—polar bears that have a visual appearance, door knobs that you can physically turn, and rock classics that actually sound like something. But this only scratches the surface of what we can talk about. One of the unique and powerful things about human language is that we can use it to talk not just about the easy, concrete stuff but also about ideas that we can't see or feel. We can talk meaningfully about *truth, responsibility*, or *justice*, none of which really look like anything. Or, for that matter, we can talk about meaning itself, like this book does. (How meta is that?) Somehow we're able to make meaning about meaning. How do we do it?

The question of how we understand abstract concepts like these is really important, because at first blush it seems like a huge stumbling block for the embodied simulation hypothesis. If simulation of sights, sounds, and actions is really at the heart of meaning, then how could we ever understand language about things that we can't see or do? For example, how do you manage to grasp the meaning of language about understanding, like this very sentence?

It might seem like an insurmountable hurdle—after all, what would it be like to simulate abstract concepts? But there are actually some clues scattered about. The first one comes from the very language we use to talk about abstract things. Take a look at how I started this chapter. I said I've been *clubbing you over the head with study after study*. I said that you might have *swallowed that idea* and asked how you *grasp the meaning of language about understanding*. Now, with those words, I clearly wasn't claiming to have literally clubbed you, any more than I was talking about actual swallowing or grasping. When I said I'd been *clubbing you over the head*, I was really saying that I'd been describing lots of evidence in order to overwhelmingly convince you. So I was using language about a very concrete action (*clubbing*) that looks and feels like something quite vivid to describe something much less tangible (convincing). In other words, I was speaking metaphorically.[1] And metaphorical language provides a hint about how we might be able to understand language about things that don't look like anything. Maybe we don't just use words for concrete things like clubbing and grasping to talk about abstract concepts like convincing and understanding. Maybe that's also the way we think about them.

Understanding, Metaphorically

Let's dig a little deeper into how metaphorical language works. How do we actually talk about intangible notions like *truth*, *love*, *values*, *irrational numbers*, or *society*? Let's begin with that last one, *society*, which is about as abstract a concept as there is. I have no idea what society looks like, what it smells like, or what it sounds like, and I'd wager you don't either. Yet, we're able to say meaningful things about it. Here are some examples of things people have said about society, all of them taken verbatim from actual web pages:

White's Residential & Family Services, . . . serves more than 3,400 at-risk children and families annually—those who without assistance and guidance would fall through the cracks of society.[2]

Japan has been a closed society for long despite its huge outward economic expansion with the world.[3]

War veterans struggle to fit back into society.[4]

Looking closely, you'll see that the way these sentences talk about society is a lot like the way we talk about concrete things. And not just any concrete things but some particular concrete things—namely, concrete things that can have *cracks* that stuff can *fall through,* that can be *closed,* and that people can *fit into.* What kind of things has these properties? It seems like people talk about society as though it were a sort of container, one that holds society's members in it. Of course, there are other ways to talk about society—take a look at this other set of examples, also all actually attested:

Farmers are the backbone of our society.[5]

Sexual violence disempowers women and cripples society.[6]

A healthy society requires an ongoing dialogue between faith and reason.[7]

The way society is described here is a strikingly different from what we saw before. Now society is something that has a *backbone,* can be *crippled,* and can be *healthy* or not. It's not so much a container as an organism. So there are at least two different patterns at play here.

All this is metaphorical, because society isn't actually either of these more concrete things—it isn't really a container or an animal. Now, before you start worrying about the fact that these sentences don't have the canonical form of metaphor you learned about in school, *X is* Y, as in *my love is a red, red rose*, let me ease your mind. Even though they don't look like the metaphors you're used to, they're still metaphorical. Any time you have language that normally describes a concrete thing (like a container or an organism) being used systematically to describe some other, abstract thing (like society), you're looking at metaphor. This just isn't your high school English teacher's metaphor.

The chief benefit to looking at metaphorical language is that it gives us a hint about how abstract concepts might be understood through embodied simulation. When using metaphor, you describe abstract concepts in terms of concrete ones. Perhaps—and this is a big hypothesis—you also understood them in those terms. In other words, when you use metaphorical language like *War veterans struggle to fit back into society*, which describes veterans trying to physically fit into something like a container, maybe you actually mentally simulate the concrete, physical motion it describes. Maybe you understand abstract concepts through concrete, though metaphorical, simulation. Let's call this the *metaphorical simulation* hypothesis.

If this is true, it's a very big deal. One of the most imposing potential weaknesses of the embodied simulation hypothesis—abstract concepts—could be dealt with using the same embodied simulations that people perform to understand concrete concepts. Maybe simulation goes way deeper than we ever thought. Maybe the greatest weakness of embodied simulation is actually its greatest strength.

So how would we begin to investigate this new metaphorical simulation hypothesis—how could we tell whether you understand language about abstract concepts through concrete simulations? For starters, we'd need to figure out what concrete things people are describing abstract things in terms of, by looking at metaphorical language. And when you start to look for it, you find metaphorical language everywhere. Just look at the preceding paragraphs. A *big deal* isn't literally big. Embodied simulation can't literally *go deeper than we thought*. The problems posed by abstract concepts can't literally be an *insurmountable hurdle*. So there's lots of metaphorical language to choose from. Where to begin? Some of the best studied metaphorical language is in metaphorical idioms.[8] An idiom is a conventional expression where a collection of words are used with a specific, idiosyncratic meaning, like *you scratch my back, I'll scratch yours* or *the more the merrier.* Some idioms additionally have metaphor built into them. For example, think about expressions like *spill the beans* or *swallow your pride.* In both of these idioms, the language metaphorically describes abstract things (secrets or pride) as though they were physical objects that can be spilled or swallowed.

So once we've found some clear examples of metaphorical language, the next thing we want to know is whether people actually simulate these metaphorically described objects and actions. So do you actually simulate swallowing something when you hear a sentence about *swallowing your pride*? Likewise, do you simulate beans pouring out of a container when you hear a sentence about *spilling the beans*? That's what our metaphorical simulation hypothesis would predict. Perhaps you interpret metaphorical language quasi-literally—rendering the concrete things and actions it describes in embodied simulations that you use in understanding the actual abstract concepts.

This idea has been tested in a number of ways, but probably the most direct comes from some work done by psychologists at U.C. Santa Cruz.[9] People came into the lab and, before the actual experiment began, were first trained to perform specific actions when prompted by symbols. For instance, if they saw " on the screen, they were supposed to make a grasping motion with their hand. If they saw #, they were supposed to swallow. And so on. Then came the main experiment. In it, they would first see one of these symbols that they had learned, and were supposed to perform the appropriate action. Then a metaphorical phrase appeared on the screen. And critically, it either matched the action (e.g., the person had just grasped and then saw *Grasp a concept*) or mismatched it (e.g., the person had just grasped and then heard *Swallow your pride*). The participants were supposed to push the space bar as fast as possible once they had understood the phrase. If understanding a phrase uses embodied simulation of actual grasping or actual swallowing, then having just performed that same action should lead people to understand the metaphorical language faster.

The results showed that people were about half a second faster to understand the metaphorical phrases when they matched the action they had just performed. That is, it took people less time to decide that *grasp a concept* made sense in English right after they performed an actual grasping action than after they performed a swallowing action. But before we jump to conclusions (that's another metaphorical idiom, by the way), it's important to note that this result could actually mean one of two very different things. It could be that performing an action makes you faster to understand a phrase that uses a metaphorical description of the same action. But on the other hand, it could be that performing an action makes you slower to understand language about a different metaphorical action. So, to tease these possibilities apart, the experimenters also included a condition where the participants saw a blank box before the sentence, which told them that they were supposed to perform no action at all. They found that this baseline condition led to the same reaction times as the mismatching condition, but significantly slower responses than the matching condition. The only condition that was different was when the action people performed was the same as the action they subsequently performed. From this, we can infer that performing an action speeds up understanding of a subsequent matching metaphorical phrase.

There are several ways to interpret this result, but the one most consistent with the work we've looked at in the preceding chapters is that understanding a metaphorical action phrase, like *grasp a concept*, activates the motor apparatus responsible for performing the same action; in other words, mentally simulating the metaphorical action. This would explain why first performing an action speeds subsequent comprehension of a matching metaphorical phrase. Performing the action warms up your motor system so that it's easier to use it in the same way when you're subsequently understanding metaphorical language about that same action.

If this account is right, then people's comprehension of metaphorically compatible language should also speed up even when they don't have to perform actions but instead just imagine performing them—either way, the resulting body part–specific motor system activation should facilitate subsequent, compatible metaphor processing. To test this hypothesis, the same authors conducted a second experiment, identical to the first, except that, when participants saw an action symbol (like #), they were supposed to merely imagine performing the assigned action, instead of actually enacting it. And sure enough, they got the same results. Imagined matching actions produced faster metaphorical phrase comprehension than mismatching ones did, or than imagining no action at all. So it appears that merely simulating an action makes you understand a phrase faster if it metaphorically uses that same action. And this suggests that understanding metaphorical language involves embodied simulations of the specific details of the metaphorical language, like the concrete actions that it describes abstract activities in terms of.

Another way to investigate how metaphorical language links into systems for perception and for action relates to words that have very similar meanings. Consider the words *joy* and *happiness*. Do they mean exactly the same thing? If not, what's different about them? It turns out that, although you probably aren't aware of it, one of the differences is in the metaphors we use for these two words. We're more likely to talk about *happiness* as an object that we seek out and acquire; for instance, we might say that we're *searching for happiness* or *sharing happiness*. On the other hand, *joy* is more likely to be described as a liquid where we're a container for the emotion, as in *I'm filled with joy* or *He's overflowing with joy*. These differences are subtle, and they're only statistical tendencies—we can say we're *searching for joy* or *full of happiness* if we really want to, even though we're less likely to do so. And if you don't believe me, look at the chart in Figure 54. These numbers are from a study that looked at a large collection of written and spoken English from different sources and counted how often each of these words was used with each metaphor.[10] People use *joy* almost twice as often with the container metaphor as with the searching metaphor, but

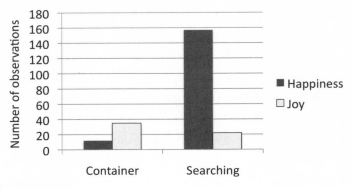

FIGURE 54

conversely they use *happiness* almost ten times more often with the searching metaphor than the container metaphor.

But how deep does this difference go? Does this mean that we also understand the meanings of the words differently? Do we perform different embodied simulations when thinking about *joy* versus *happiness*? One way to test this idea would be to measure whether people are more likely to use the word *joy* when they are in fact actually being a container for liquid, and, by the same token, whether they're more likely to use the word *happiness* when they're actively searching for something. If they are, this would suggest that the meanings of the *joy* and *happiness* are in fact linked to the body and brain systems that are responsible for experiencing containment and search, respectively.

We tested this idea in the following way.[11] We printed out the picture in Figure 55 with the question underneath it.

FIGURE 55

What emotion is this person experiencing?
A) JOY
B) HAPPINESS

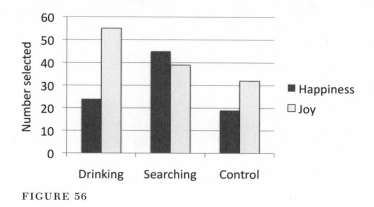

FIGURE 56

And then three graduate students working with me went around and asked people to answer the question. The trick was that they asked three different groups of people. The first group was actively engaged in filling their bodies with liquid—that is, they were drinking—at cafés, bars, and so on. The second group was actively searching for something, like a book in the library. And the third group, the control group, was neither drinking nor searching; they were just sitting in a classroom. What we found was that people who were drinking were far more likely to say that the person was experiencing *joy*. The searchers were more likely to say *happiness*. And people who were doing neither fell somewhere in between the other two. You can see the results in Figure 56.

The upshot is that the current state of people's bodies affected the words they used. This is again compatible with the idea that the meanings of relatively abstract words are based in embodied simulations of the more concrete things they're described in terms of.

Metaphor in the Brain

Of course, if people are really constructing concrete simulations while understanding metaphorical language, then we should see some indication of this in the brain. Several quite recent studies have asked, using brain imaging, whether we use our motor systems to understand language about *grasping ideas* or *clubbing you over the head with study after study*. The first study to directly ask this question placed people in an fMRI scanner and presented them with either literal or metaphorical action sentences, like *Bite the apple* or *Bite the bullet*.[12] They then had the same people watch the same physical actions in the scanner, so they could tell exactly where the parts of their brains were that recognized those actions and whether those same parts were also active while people were processing the sentences. But what they found was that the metaphorical language describing actions using different body parts did not lead to activation in the relevant locations in the motor system. Another study subsequently looked at the same question in a very similar way.[13] They had people move the various parts of their bodies in the scanner to localize their motor areas, had them separately understand literal or metaphorical sentences, and came up with a very similar outcome—metaphorical language didn't light up motor areas. So is the metaphorical simulation hypothesis dead on arrival?

It would seem like it. But other brain imaging studies have turned up results that are much more compatible with the metaphorical simulation hypothesis. One study used a very similar method to the first two, with a couple differences.[14] It, too, used metaphorical idioms like *kick the bucket* but used a whole lot more of them to see whether the missing simulation effect from the earlier studies could be due to just not having enough data. And, in addition, it changed the method so that instead of presenting a whole sentence in written form all at once, they presented sentences, still visually, but one word at a time. This meant that although people in the experiment were reading, they were doing so in a way that was a little bit more like listening to spoken language, because they were getting the sentence incrementally. With just these two differences, all of a sudden, parts of the motor system started to light up when people were reading about metaphorical actions. And not only that, but regions that control actions with specific body parts were more active when the metaphorical language was about actions using those body parts.

So, what gives? Why would it be that language about metaphorical actions sometimes does and sometimes does not activate the motor system? One possible explanation is that the one study that did show a metaphorical simulation effect presented sentences one word at time. It might just be that when you're processing metaphorical language, part of the reason you activate your motor system is because you get tricked into thinking that the language is actually literal—you think you're dealing with a sentence about actual kicking, so you start simulating kicking, and then all of a sudden the rug gets pulled out from under you and you're staring straight at a metaphorical idiom.

There's another possible explanation: the number of sentences in the studies. Maybe the use of simulation for metaphorical language is just less pronounced than it is for literal language, such that it takes more data for it to pop out. And there's some circumstantial evidence for this explanation, as well. When you look at the type of metaphorical language used in all these studies, they were all metaphorical idioms—those memorized chunks of language that, by convention, have a particular use. So people were reading sentences like *Bite the bullet* and *Kick the bucket*. Maybe those metaphorical expressions, by dint of their long careers in the language, have come to be bleached of their literal meaning.[15] By comparison, more innovative metaphorical language, like *Bite into this idea* or *Kick this meeting into overdrive* might encourage relatively more residual motor simulation.

So a more recent brain imaging study decided to pursue this second idea and looked at whether idioms are less likely to engage the motor system than less familiar metaphorical expressions.[16] They again placed people in an fMRI scanner and had them read sentences about literal or metaphorical actions. Some of those sentences used more familiar metaphorical expressions, and some used less familiar ones. The way they presented the sentences was a sort of hybrid of the preceding studies. Sentences were broken into two parts—first the subject was shown and then the predicate—so people might first see *The public*, and then after half a second they would see *grasped the idea*. So this again made reading a little more like listening to language, in that it broke it up incrementally, but it specifically avoided the possibility that people were being led down a garden path by the verb. What they found was that, first of all, both types of metaphorical expression—the more and less familiar ones—activated the participants' motor systems. But, more intriguingly, the more familiar the expression, the less it activated the motor system. In other words, over their careers, metaphorical expressions come to be less and less vivid, less vibrant, at least as measured by how much they drive metaphorical simulations.

So what does the brain tell us? First off, the metaphorical simulation hypothesis isn't straightforwardly overwhelmingly right. There seems to be some language, language that appears metaphorical like familiar metaphorical idioms that, when you read it as whole sentences, doesn't always massively activate the relevant parts of the brain that we might expect to light up if people are performing motor simulations. Secondly, under other conditions, for instance, when the metaphorical language is less familiar or when it's presented incrementally, all of a sudden things start to fall in line with the metaphorical simulations we might expect. In sum, there's something happening here, but what it is ain't exactly clear.

What Are Metaphorical Simulations Like?

So far, we've seen some behavioral evidence and some brain imaging evidence that people do at least sometimes perform embodied simulations when they're interpreting metaphorical language. This offers a tantalizing possibility—that perhaps we understand language about abstract things, like society and joy, similarly to how we understand language about concrete things.

But, at the same time, this idea only raises more questions. We know that when we hear metaphorical language, like *John grasped the idea*, we often seem to engage our motor system to simulate an action. But exactly what is that simulation like? An idea isn't a physical object, so when we simulate grasping an idea, what exactly is it that we simulate grasping? Is it some physical object that stands in for ideas? Or do we somehow simulate merely grasping, without any specific object in mind? And, likewise, when we describe abstract concepts in visual metaphorical terms, like *That idea is becoming clearer*, do we perform visual simulation of a specific object becoming clearer? If so, what does it look like, because ideas don't look like anything! Or is there a way to simulate something getting clearer without it actually looking like any particular object? You see the issue.

One experimental setup we've already seen actually gives us some clues about exactly how specific metaphorical simulations are. We saw an experiment earlier in which people heard sentences that described upward or downward motion (*The mule climbed* versus *The chair toppled*), and, as you might recall, this interfered with their ability to perceive a shape that appeared in the same region on the screen. But what would happen if we used the same verbs, like *climb* and *topple*, but made them metaphorical, by pairing them with abstract nouns? That would give us sentences like *The rates climbed* or *The prices toppled*. If we use our vision system to mentally simulate objects moving in one direction or another while processing metaphorical sentences like these, then these metaphorical uses of the same verbs should show the same interference effect that the literal uses did. Well, we did that experiment in our lab, and it turns out that what happens is nothing—there's no interference effect.[17] A sentence describing metaphorical motion upward, like *The rates climbed*, doesn't interfere with perceiving a shape, no matter where it appears on the screen.

What's interesting is that this experiment shows no indication of visual simulation while people are understanding metaphorical sentences, although other research on metaphorical language comprehension we discussed does. Why don't the different methods converge on the same answer? It's possible that the differences between the experimental designs actually tell us something very important about the level of specificity of people's metaphorical simulations. The first two experiments we saw in this chapter—the action-miming and joy-happiness experiments—did not require people to perceive or interact with any specific object. And the results from those studies showed

that people were performing embodied simulations while understanding metaphorical language. But the Perky method we used with
The rates climbed depends on location-specific interference in the visual system between perceiving a particular object and mentally simulating a different object in the same location. And in this method,
there was no indication of simulation. The difference between these
methods is whether the simulations we were measuring included objects or not.

So it could be that metaphorical language like *The rates climbed* or
Grasp an idea involve embodied simulation of upward motion or of
grasping, but that those metaphorical simulations are less detailed
than simulations for literal language because they don't actually include a physical object. And this would explain why the Perky method
shows no effect for metaphorical language—if there's no specific object being mentally simulated, then there's no image to interfere with
actual perception. On the other hand, the other two metaphor studies
(the action miming one and the *joy/happiness* one) merely measured
whether performing an action would prime the recognition or use of
a particular piece of language. There's no specific object in these studies to be compatible or incompatible with—in the grasping case, for
instance, the participants were merely enacting a grasping action without an object. And, as a result, this may have been more compatible
with embodied simulations appropriate to the metaphorical language
people were understanding.

So the long and the short of it (so to speak) is that to understand
metaphorical language, we appear to construct embodied simulations
that are slightly less detailed than ones we construct for literal language but that are no less motor or perceptual. There's something
very exciting about this finding. We already knew that embodied simulation is active when we're understanding language about concrete
things. But that only skims the surface of what we can do with language. So much of what we actually talk about is abstract that we
could hardly say we understand the process of understanding without figuring out how people grasp abstract concepts. The idea that

we've come to is that we take what we know about how to perceive concrete things and to perform actions, and we use that knowledge to both describe and also think about abstract concepts. In this way, we bootstrap harder things to think and talk about—abstract concepts— off of easier things to think and talk about—concrete concepts.

Speaking Abstractly

If you thought it was tough to figure out how people understand metaphorical language, then hold on to your hat. Because at least metaphorical language gives you a clear indicator of what concrete things to mentally simulate. The metaphorical idiom *hold on to your hat*, to take a salient example, tells us what to simulate doing (holding) and what to simulate doing it to (our hat). But language about abstract concepts isn't always this helpful. A lot of language about abstract concepts doesn't actually specify anything concrete to think in terms of. For instance, consider the difference between the metaphorical sentence *The prices toppled* and the abstract sentence *The prices decreased*. It's subtle, but the difference is that the verb *topple* tells us to simulate downward motion, while the verb *decrease* doesn't tell us anything about movement in space. *Decrease* doesn't have a concrete spatial meaning like *topple* does. If you say *The leaves on the tree decreased*, that doesn't mean they moved downwards.

So how do we understand abstract language that's not explicitly metaphorical, like *The prices decreased* or *I believe in social justice*? This is a hard problem for the embodied simulation hypothesis—even harder than metaphorical language is. But it's no less important. So let's see how embodied simulation might work for abstract language. The most obvious possibility is that we understand abstract concepts through simulating different concrete things, just like we do during metaphor, even if the connection between abstract and concrete isn't stated explicitly in a given sentence. For instance, we might understand *The prices decreased* in the same way that we do *The prices toppled*—by mentally simulating downward motion. In other words, we might build metaphorical simulations for language about abstract concepts even when the language that we're dealing with isn't itself metaphorical.

But how would we know what to metaphorically simulate? Abstract language doesn't specify what concrete domain to engage during simulation—that's what makes it abstract and not metaphorical. And as we've seen, abstract concepts like prices and society can actually be cast metaphorically in terms of lots of different concrete, simulatable things—society can be a container or an organism, or a variety of other things. This would lead us to predict that people might actually vary a lot in what concrete domain they metaphorically simulate these abstract concepts in terms of. So, although I might simulate prices decreasing in terms of downward motion, you might simulate it differently, say as a decrease in spatial extent, like you would for the metaphorical sentence *The prices retracted*.

Fortunately, there's been a little bit of work looking at how we understand abstract language. One study had people perform the now-familiar action-sentence compatibility experiments (where people read a sentence then perform a compatible or incompatible action), with abstract rather than concrete sentences.[18] The researchers needed to find abstract language that might be understood in terms of motion toward or away from the body, even though it wasn't itself metaphorical. What they came up with was language about transfers of information, like *You radioed the message to the policeman* or *The policeman radioed the message to you*. These sentences don't use concrete spatial language like *give* or *receive*, so they're not metaphorical. But we do know that, in other contexts, we talk metaphorically about communicating in terms of giving and receiving, as in *Let me give you an idea* and *I got this great idea from her.* So it's plausible to think that we might also think about communication in terms of giving and receiving. And sure enough, when they stuck abstract sentences like the ones about *radioing* into the action-sentence compatibility method, they found an effect that was just as strong as the one produced by the concrete sentences.

This metaphorical simulation story is appealing, because it claims that we understand language about abstract concepts in the same way, regardless of whether the language in question is metaphorical. But it can't be the whole story. And this is for a reason that actually applies to metaphorical language as well. Suppose that when we understand language about abstract concepts, we perform concrete simulations that are akin to the ones we perform when we process language about concrete concepts. If these simulations play a role in language comprehension, contributing in some way to making appropriate inferences, updating our beliefs appropriately, and so on, then there has to be some difference between simulations for concrete concepts and the ones for abstract concepts. After all, when you hear *The prices decreased* or *The prices dropped*, you clearly don't get confused into thinking that there's some physical thing that actually fell. So there must be something different in the simulations we perform for concrete and abstract concepts, or in the way that we use those simulations. Simulations that people perform for abstract concepts do appear to be less specific, at least in terms of objects, than ones for concrete language. But just being less specific isn't enough to keep us from getting confused into believing that prices actually literally went downwards. As of the writing of this book, we don't have an answer to this problem. We don't know how metaphorical simulations differ from literal ones in such a way that we're able to keep our simulations straight. But that's how science works. The better your answers get, the more questions they produce.

Metaphor Without Language?

Research on metaphorical language has led to lot of interest in the idea that people not only talk about abstract concepts in terms of concrete ones but might also <u>think</u> about them in those terms. In a way, the strongest test of this idea would be to see whether people are using concrete things to understand abstract concepts when they're not using metaphorical or abstract language at all. And in recent years, there's been a groundswell of research testing for exactly this.

I'll start with a very compelling study but one that's also pretty complicated. So buckle your seatbelt. This study looks at how we think about time.[19] Time is obviously a quite abstract thing—it doesn't look or feel like anything. The way we talk about time (at least, one very common way) is in terms of space. Because, unlike time, space is concrete—we can see distances and changes in location, we can explore space by moving through it, and so on. And the two go together. When we measure time, we often do so by relying on space—how far the little hand has moved on the clock or how far the sun has moved in the sky. And our language reflects this as well. We talk a lot about time in terms of space, but the reverse is quite rare. Here is some metaphorical language, describing time as space:

Spring break is still way ahead of us.

Two hours is way too short for me to finish this exam; I need a longer time.

My sister's birthday is so close to Christmas that she gets like half the presents that the rest of us do.

But the reverse isn't true—we don't talk about space in terms of time. It would be weird to say *That blue station wagon is still in the future*, meaning that the station wagon is spatially ahead of you. And it would be equally strange to say *If you get lost in the mall, stay at the same time*, meaning that you should stay in the same place. We talk about durations in terms of distances, but not distances in terms of durations.[20]

The question addressed by researchers at Stanford was whether people not only talk about time in terms of space but also think about time in terms of space, even when there's no language involved. Here's the experiment they designed to get an answer. People saw a line on a computer screen that grew from left to right for a variable amount of time and for a variable distance. Specifically, it could go any one of nine distances, and could take any one of nine durations of time to do so. But the crucial thing was that there was no relation between the time that the line took and the distance it traveled. That is, for each participant, the line went each of the nine distances in each of the nine durations of time. That's an important point, so let me underline it: <u>there was no relation between the length of the line and the amount of time it took to appear</u>. You couldn't use one to determine the other.

What participants had to do was to watch the line each time it grew on the screen and then make an estimate about it. There were two things they might have to estimate, either time or distance. The way they knew which estimate to give was that after the line was done, they'd see one of two prompts. If they saw an hourglass, this told them to estimate either how much time the line took (by clicking twice with a mouse, once for the beginning and once for the end). And if they saw an X, then they were supposed to estimate how far it went (by clicking on the starting and end-points on the screen). The question was whether the spatial length of the lines would affect people's time estimates or the reverse or both.

The results were exactly in line with the metaphorical language people use. When people were asked to estimate the amount of time that the line had been on the screen, they were strongly affected by its spatial length. Their average time estimates correlated almost perfectly with the distance the line had traveled; the farther the line had gone, the more time people thought it had taken. This is important, because, as you might remember (because I underlined it!) this correlation wasn't a property of the stimuli—there was no relation at all between the actual distance the line went and the actual time it lasted for. So people were mistakenly using distance to estimate time, despite the fact that this made their answers less correct. You can see this in Figure 57 on the left. Correct responses would result in the horizontal dashed line across the chart, where distance (on the x-axis) had no effect on estimated duration (on the y-axis). Let's translate this into a situation that might be a little more intuitive to you. If you're watching people compete at the long jump, your perception of how long it's taking them to jump will depend almost entirely on how far they jumped. The farther the jump, the more time you'll think it took.

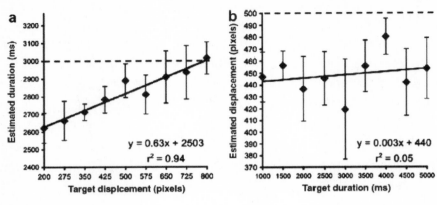

FIGURE 57

But what about the reverse? Does time affect distance estimates? In Figure 57 on the right, you see what happened when people made distance estimates. There, the line is basically horizontal, indicating that people's distance estimates—how far they thought the line had gone—did not correlate at all with the duration of the line in time. (In long jump terms, you don't think people went farther just because they took longer.) In other words, even when there's no language around (just lines on screens), people use space to make judgments about time but not the reverse. This adds support to the idea that abstract concepts are generally understood in terms of more concrete ones, and not the reverse, even when there's no language to prompt them to do so.

Since 2005, there's been a cavalcade of other studies showing similar things. For example, we talk about affection as warmth—contrast a *warm smile* with an *icy stare*. One ingenious study manipulated how physically warm people were—by surreptitiously asking them to hold a cup of warm or cold coffee—and then having them make a judgment about how friendly an imaginary person was.[21] They found that people who were given physically warm cups of coffee thought that the imaginary person was more generous, happy, and sociable—attributes that are described in terms of warmth—but not stronger or more honest, which are not described in terms of warmth. Another study asked people to think about times they felt either socially included or excluded. Then, very cleverly, the experimenters told the participants that they were having trouble with the central air in the building and asked them what they thought the temperature in the room was. The participants guessed that the temperature of the room was significantly colder when they were thinking about feeling socially excluded than when they felt socially included.[22] In other words, lonely feels cold.

Here's another example. We talk about morality in terms of cleanliness—for example, if you commit *dirty deeds* you'll *sully* your otherwise *clean record*. People also think about morality in terms of cleanliness. If you have people recall either an ethical or unethical deed from their past and then offer them a choice of parting gift for participating in the study—a pencil or a cleansing wipe—the people who were asked to recall unethical actions are three times as likely to pick the cleansing wipe, while people asked to recall ethical actions are twice as likely to pick the pencil.[23] In other words, when people think about unethical actions, they seem to feel a need to cleanse themselves. This has been dubbed the *Macbeth effect*. But the most amazing aspect of this effect is that cleansing actually appears to help people "clean away" moral transgressions. One study had all the participants recall an unethical deed from their past. Half were then instructed to disinfect their hands with a cleansing wipe, and half were not. Then they were asked whether they'd be willing to volunteer, without pay, for another experiment—in other words, to do a favor for someone. Remarkably, the people who <u>did</u> cleanse their hands were about <u>half</u> as likely to volunteer for the experiment as the people who did not cleanse their hands. Possibly, this was because they felt that they had washed away their transgressions and didn't need to do good to feel better about themselves. But those participants who didn't have a chance to clean their hands still felt immoral and as a result felt they needed to rectify the situation by doing something good for someone.

What does all this new research show? At the very least, it shows that abstract concepts like time, morality, and affection are tightly linked to the very concrete things that they're metaphorically described in terms of—distance, cleanliness, and warmth. And this is the case even when there's no metaphorical language present, which would seem to go along with the metaphorical simulation view of how we deal with abstract concepts. Although research on how people understand abstract language is still in its infancy, evidence like this suggests that abstract concepts are understood in terms of concrete ones, whether during language processing or on their own.

Language on the Move

One of the neat things about the metaphorical simulation hypothesis is the idea that we might be simulating one thing to understand another. There's actually another situation that this might apply to. But it doesn't relate abstract concepts to concrete simulation, like metaphorical language does. Instead, it has to do with two different types of concrete things. Let me start you off with an example. Suppose you're standing on your veranda, gazing at the path that runs from the stoop in front of you down a small hill to the street at the bottom of your property. To describe this path, you might say something very much like what I just wrote: the path *runs down the hill*. Now, a moment of reflection. Why would you use the word *run*? After all, there's no actual running involved when you're just looking at the path or describing it. When you first think about it, you might suspect that you use *run* because someone might run on the path. That's possible, but there's good reason to think that that's not exactly what's going on, because other things that you can't run on—like *scratches* and *cracks*—can nevertheless *run*; and, what's more, they can *meander*, *zigzag*, or *climb*:

The used monitor has a scratch that runs all the way across the top.

The river meanders down through the foothills.

The crack zigzags across the wall of the garage.

*The gutter climbs to a relatively high apex in the exact middle of
the house.*

When we take language that normally describes motion through
space and use it to instead describe how things are arranged stati-
cally in space, like these examples do, it's called *fictive motion*. And
the reason I'm bringing it up here is that fictive motion might be
like metaphorical language in that it evokes simulation of one thing
when we're actually understanding something else. In particular,
perhaps using fictive motion inserts motion into the embodied simulations we run, even though the actual scenes being described have
no motion in them. So the difference between *The highway runs
parallel to the river*, which uses fictive motion through the verb *run*,
and *The highway is parallel to the river*, which does not because of
the static verb *is*, might be that you simulate motion when you
process the first sentence, but you simulate a static spatial configuration for the second.

If this intuition is right, then there should be ways to test this empirically. For instance, longer paths or ones that are harder to traverse
and thus must be navigated more slowly, might take longer to mentally
simulate than shorter, easier paths. Teenie Matlock, a cognitive scientist at U.C. Merced, tested this in the following way.[24] First, she had
people read a series of sentences describing a path covering either a
short distance or a long distance. For instance, below are comparable
short and long distance scenarios:

Short Distance Scenario:
Imagine a desert. From above, the desert looks round. The desert
is small. It is only thirty miles in diameter. There is a road in the
desert. It is called Road 49. It starts at the north end of the desert.
It ends at the south end of the desert. Maria lives in a town on the
north end of the desert. Her aunt lives in a town on the south end.
Road 49 connects the two towns. Today Maria is driving to her
aunt's house. She is driving on Road 49. It takes her only twenty

minutes to get to her aunt's house. After she arrives, Maria says, "What a quick drive!"

Long Distance Scenario:
Imagine a desert. From above, the desert looks round. The desert is large. It is four hundred miles in diameter. There is a road in the desert. It is called Road 49. Road 49 starts at the north end of the desert. Road 49 ends at the south end of the desert. Maria lives in a town on the north end of the desert. Her aunt lives in a town on the south end. Road 49 connects the two towns. Today Maria is driving to her aunt's house. She is driving on Road 49. It takes her over seven hours to get to her aunt's house. After she arrives, Maria says, "What a long drive!"

And then, immediately after each scenario, participants saw a fictive motion sentence like the following and were asked to decide whether it related to the preceding story or not:

Road 49 crosses the desert.

What she found was that people took almost half a second longer to respond "yes" to the fictive motion sentence (that is, to indicate that the sentence was related to the previous story) when the paragraph described long-distance motion than when it described short-distance motion. This implies that they were performing longer simulations when the language described motion that would take longer to actually perform or observe—exactly in line with what you would expect if people were simulating actual motion when processing fictive motion sentences.

But there's an alternate explanation of this finding. It could be that the long-distance scenarios merely slowed down people's processing in general. Maybe they lulled participants into slower reactions to the next task, regardless of what that task was. And maybe the short-distance scenarios got people amped up so they would perform any subsequent task more quickly. To test this, Matlock replaced the fictive motion sentences (*Road 49 crosses the desert*) with literal, static ones (like *Road 49 is in the desert*). If people were just acting lethargic in general after the long-distance scenarios but energetic after the short-distance ones, then they should show the same difference in reaction time when asked whether these new sentences were related to the previous story (faster responses after the short story, slower after the long story). But it turned out that it didn't matter whether people had just read the long or the short scenario. How long it takes to process a fictive motion sentence reflects how far that fictive motion would go, but, for static sentences, there's no such indication of motion simulation. And, what's more, Matlock subsequently replicated the same results for scenarios that differed in the speed of travel and difficulty of the terrain.

So it appears that understanding a fictive motion sentence leads you to simulate motion. But motion of what sort? There are a couple possibilities. On the one hand, we could be imagining an object moving at a particular rate or over a particular distance. That seems like it might work for fictive motion sentences about paths, like roads and the like. But it's more of a stretch for non-path-like things, like scratches and roofs. An alternate possibility, one that could account for fictive motion in either paths or nonpaths, is that we're not simulating the scene as moving; instead, we're simulating moving our gaze through the scene. In other words, if you read that *The river meanders down through the foothills*, maybe you imagine a visual scene and move your visual attention across this scene along the path that the fictive motion sentence describes—in this case, meandering down through the foothills. Maybe the thing moving is your simulated focus of attention. This second account seems more promising because it could straightforwardly account for both types of fictive motion—motion over paths (like roads) and nonpaths (like cracks).

And if it's right, this second idea suggests that fictive motion language is a lot like language about how we move our gaze. When we talk about looking at the world, we often use language that metaphorically describes the eyes or the gaze as touching the objects we're looking at. We talk about *making eye contact* or *our gaze alighting on someone's face* or *a piercing stare*. If focusing on something is metaphorically touching it, then shifting your focus is metaphorically like moving your gaze through the world so that it touches different things. It could be that fictive motion is akin to this type of metaphorical language.

Matlock joined up with eye-tracking researcher Daniel Richardson to design a way to test this.[25] What they did was to have people look at images of objects, for instance, books on a shelf, like in the image in Figure 58. And, at the same time, the participants heard a sentence, which was either a fictive motion sentence, like *The books run along the wall*, or a static sentence, like *The books are on the wall*. Participants were supposed to determine whether the sentence described the picture. And, all the while, the researchers were using an eye tracker to see where the participants were looking. What they found was that people spent more time looking along the axis of the things described as moving, that is, the books, when the sentence was a fictive motion sentence than when it was a static description.

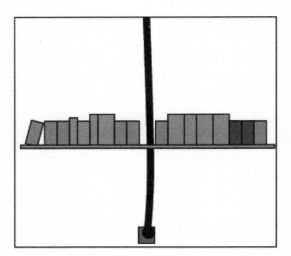

FIGURE 58

Actual eye movements are a good indicator of mental attention, and this experiment suggests that people understand fictive motion language about things that aren't actually moving at all by moving their mind's eye over simulated scenes. Again, this is a way that people appear to be simulating one thing (the books aren't literally *running*, after all) to understand language about something quite different.

Lots of Open Windows

My high school English teacher used to say that when you conclude a paper, you should always *close a door and open a window*; the door you're closing is the issue at hand, which you should have resolved, and the window is the set of questions you've raised and prospects for future research. This advice is quite reasonable (and notice that's it metaphorical as well—you did simulate a door and a window, right?). But the conclusion to this chapter will unfortunately have to involve less door closing and more window opening than my high school English teacher would have liked. That's because abstract concepts are hard. The research described in this chapter takes a look through the doorway—we know one way that language about abstract concepts might be understood and have seen some evidence about this proposed account. But, at the same time, there are still some very hard questions to answer.

For instance, if metaphorical and abstract simulations are similar to ones performed when people process concrete language, how is it that people don't get confused by the simulations they perform when understanding metaphorical language—why is it that they don't believe that *John grasped the idea* is actually about physical grasping? You know that my English teacher's advice about the door and the window isn't actually about doors and windows. So it can't be the case that the cognitive processes that underlie understanding abstract concepts are identical to the ones underlying concrete ones. Therefore, the more that understanding abstract language is like understanding concrete language, the more infrastructure we must have to keep them apart.

How abstract and concrete concepts are different, how embodied simulations relate to the inferences we make from them, and how we use simulation to update our beliefs about the world, are all big questions that are up in the air. They're most definitely worth opening a window for.

CHAPTER 10

What Is Simulation Good For?

It's time to take stock and ask some hard questions. Over nine chapters, I've laid out what I hope is a pretty convincing and coherent account of how detailed, variable, and pervasive embodied simulation is in language use. Shortly after the sound waves of spoken words hit our ears or the light of written characters hits our eyes, we engage our vision and motor systems to recreate the nonpresent visions and actions that are described.

Let's take that as a given. In this chapter, we're going to ask some hard questions about simulation—shine the bright light of skepticism on its face and see where the wrinkles are. At issue is a fundamental question. It's all well and good to show that people perform embodied simulation when they're understanding language, but this doesn't necessarily tell us exactly what that simulation is used for. And, for that matter, it doesn't tell us even more fundamentally whether simulation serves any purpose for understanding language at all. It could very well be that your embodied simulations of polar bears, flying pigs, and yellow trucker hats, intricate and automatic though they might be, are mere ornaments that don't contribute anything to the understanding process per se. Because this is a young field, most research has focused on showing how and when what simulation occurs. There has been less concern about what that simulation is actually good for. Indeed, this remains the major criticism of this field of research that skeptics often raise. So we're going to tackle it.

A side note is in order here about skepticism. Skeptics have immeasurable value in any scientific enterprise. This isn't just because they provide a bull's-eye for intellectual target practice (although this is always useful, too). Every good scientist is a skeptic—the very premise of scientific inquiry is to not believe something unless there is compelling evidence for it and against all reasonable alternatives. The problem is that scientists aren't always skeptics in equal measure. It's often much more difficult—perhaps for personal, psychological reasons—to be skeptical of one's own work than of others'. So we often need to look to our intellectual neighbors for a healthy dose of skepticism. Skeptics are inherently useful in that they act as the measure of convincingness of a theory—if you're as good a skeptic of my theory as I am of yours, we're going to have to find more and more compelling evidence that distinguishes between them and that tells us which one is right.

So let's see how a skeptic might look at the evidence from the preceding chapters. She would probably agree (and many do) that it shows without a doubt that we perform embodied simulation when we're doing all sorts of things, including understanding language, like language about how polar bears hunt seals. She would be compelled to agree that this embodied simulation includes not only visual but also motor and auditory simulation—what the bear looks like and sounds like and maybe even what it feels like to scoot across the ice, sneaking up on a seal. And she could not help but acknowledge that this embodied simulation appears to be performed using cognitive systems otherwise dedicated to perception and action. But this is where she might stop. What is the evidence, the skeptic might ask, that this embodied simulation actually <u>does</u> anything? Can you understand language without simulating—would you still know what the polar bear was up to? Does the simulation contribute anything to the understanding you construct? Or, instead, is the embodied simulation just peripheral? Is it a mere downstream result of language understanding? Does it perhaps play no actual role—have no real function—in the understanding process?

Our skeptic might offer an alternate account of the results we've seen so far. You probably remember the Mentalese view of meaning that we saw in the first chapter. Well, the skeptic could resurrect this view in the following way. Sure, simulation happens, she would say. But suppose that language comprehension doesn't rely on simulation at all. Maybe what comprehenders do is to engage internal representations of the world, Mentalese symbols that don't have any perceptual or motor content. These Mentalese symbols are connected to knowledge about perception and action through association—when you think about kicking and activate the Mentalese symbol for that concept, you are often also perceiving or performing kicking. So, like the brain does, you've built up strong connections between the Mentalese symbols and the relevant parts of your perception and motor systems. As a result, when you activate the Mentalese symbol for *polar bear*, perhaps because you're understanding the word *polar bear*, activation spreads to parts of your vision system and you simulate what a polar bear looks like. But this apparent simulation isn't actually a part of the comprehension process per se. Instead, all the action is in the symbols. That's what the skeptic would say, anyway.

The graphical representation in Figure 59 might help illustrate the different possible explanations for the results we've seen thus far. On the one hand, it could be that embodied simulation plays a functional role in language understanding. This is what's labeled on top as the embodied simulation view. For now, we'll leave open exactly how important simulation is to language understanding. This view could see embodied simulation as <u>necessary</u> for language understanding, in which case we couldn't understand language without it. It could see embodied simulation as <u>sufficient</u> for language understanding, too, so that all we needed was to mentally simulate for understanding to be achieved. Now, because very few people would go so far as to advocate the strongest versions of these claims, for the time being, let's just characterize the embodied simulation view as arguing that embodied simulation is responsible for some aspects of normal language understanding—it is at least sometimes necessary and at least partially sufficient. In other words, it sometimes plays a functional role in language understanding. The other view characterized in the figure is a rough sketch of the Mentalese hypothesis. Here, a person processing language activates Mentalese symbols, and, on this view, selecting and activating the right set of these symbols is either necessary or sufficient (or both) for language understanding. But on this second account, embodied simulation doesn't play a functional role in the actual process of understanding (it's depicted in a dotted line to indicate that it's optional). It's a lot like the human appendix, or for that matter, male nipples—there's no question that they're there, but they don't appear to actually do anything useful.

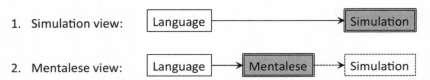

FIGURE 59 Two views of language understanding. Gray boxes indicate functional parts of the understanding process, and dotted lines indicate nonintegral processes.

To differentiate between these two theories, we will have to consider different predictions that they make. If the Mentalese view is correct—if simulation is immaterial to language understanding—then when people don't perform embodied simulation, they should understand language just as well as when they do—just as quickly and just as thoroughly. On the contrary, the simulation view predicts that if we were to interfere with embodied simulation—specifically with the aspects of embodied simulation that are thought to be responsible for processing particular sentences—then this should cause deficits in language understanding. Depending on what simulation is actually used for, when simulation is selectively interfered with, it might take people longer to come to an understanding of sentences, for example. Or they might understand sentences in less detail or less accurately. Or they might have a harder time drawing inferences or remembering the specifics. And so on.

Interfering with Simulation

Let's say, just for the sake of argument, that you're a biologist. One day your research assistant bursts into the lab with burlap sack in tow. You open it to discover—to your amazement—a flying pig! As a biologist well schooled in porcine aeronautics, you recognize immediately that it is of the Pigasus variety—it sports the compulsory pink snout, curly tail, and feathered wings. So naturally you put the pig in a comfortable pen that you appoint appropriately: with both a trough and a perch. Over the course of time, you start to wonder—you are a biologist, remember—how its delicate wings can possibly keep a pig-sized animal aloft. Perhaps, you think, the wings are mere ornaments. They look good and are presumably useful for certain things, like maybe swatting away mosquitoes, but perhaps they don't contribute to flying in any concrete way.

How would you test whether the wings play a functional role in flying? One way to do it would be to occupy the pig's wings with some other task, like swatting at mosquitoes for instance, and at the same time prompt it to fly. If having its wings busy doing something else were to interfere with the pig's ability to fly, this would certainly show that the wings play a functional role in flying. If it could still manage to fly, just not very well, then this would imply that wings are a functional part of flight, but not a necessary one. If it were totally unable to fly without wings, this would show that the wings were an entirely necessary component of flight. And of course if certain aspects of flight were impaired but not others (say, pitch control but not lift), then you'd know exactly what the wings are good for.

The same reasoning applies to language. To spell out the analogy, in case I wasn't obvious enough, embodied simulation is the wings that you want to know the functional role of. Language comprehension is flight. And you . . . well, sorry to say it, but you're the pig. Of course, the experiment is going to work a little differently. Whereas you can occupy a pig's wings by introducing mosquitoes into the mix, it's not quite as straightforward to curtail embodied simulation. One way to make sure people processing language can't use embodied simulation is to engage the visual or motor system in some other task at the same time. The basic logic is similar to the Perky effect experiments we saw in Chapter 3. If parts of the brain dedicated to certain functions (e.g., neurons responsible for perceiving objects moving toward you) are also used when people process language (e.g., about objects moving toward them), then keeping those brain structures busy with a moving visual stimulus during language processing should interfere with understanding. The difference is that where in the Perky effect experiments we saw that language-driven simulation interfered with subsequent visual perception, what we're interested in now is actually the reverse chain of interference. We want to know whether blocking simulation interferes with processing language. If language is understood more slowly or less accurately (or whatever) when the specific brain structures that perform the expected simulation are interfered with—then this implies that embodied simulation plays a functional role in the understanding process.

One simple way to test this is to simply switch the order of the stimuli in a Perky effect task, as Shane Lindsay, then a graduate student at the University of Sussex, did.[1] In each trial of his experiment, participants saw a small black rectangle on a computer screen that moved upward, downward, or horizontally for 4 seconds. Shortly after the moving rectangle disappeared (1.5 seconds later), a written sentence appeared, describing upward motion (e.g., *The diver swims to the surface*) or downward motion (e.g., *The meteorite crashes into the ground*). The participants' task was simply to read the sentence and press a button as soon as they had understood it. The prediction was an interference effect—understanders who had just observed an upward-moving rectangle, for example, should take longer to read and understand a sentence describing upward movement. If the same neural structures are used for both tasks, then engaging them for actually perceiving one thing should make it hard to also engage them for the purpose of understanding language about another thing, and as a result understanding should be measurably slower. The results bore out this prediction. It took people about 100 milliseconds longer to read a sentence describing motion in a given direction when it followed a rectangle moving in the same direction. So it seems that reading motion sentences takes longer when you are unable to use your visual system to mentally simulate the described motion.

That's cool. But it doesn't tell us by exactly what mechanism perceiving motion interfered with subsequently understanding a sentence. There are several possibilities. The first is based on the motion aftereffect that we saw earlier (remember the squirrels teeming over the fence?). Perhaps perceiving the moving rectangle led to a motion aftereffect, where motion-detecting neurons were fatigued and thus people were less easily able to mentally simulate motion in the same direction. Although a four-second exposure to a moving rectangle is quite short for an adaptation effect like this to take hold, it is possible in principle. That's one way it could have gone down.

Another potential explanation for the interference effect results from the fact that there was a pause between the time when the moving rectangle disappeared and the time when the sentence appeared. This means that the effect could have been mediated by the participants' memory systems. People might have been encoding the rectangle-moving event they observed into memory and maintaining this memory throughout the processing of the sentence. In this case, the interference would arise not because cells dedicated to detecting generic motion in one direction or the other have been habituated to a motion scene. Instead, it would be due to an event maintained in memory. Such interference of a recalled event on a very similar event engaged in by a different object would be due not to inactivity of cells dedicated to perceiving motion in a particular direction but to them already being active for another purpose.

We still don't know exactly which of these two mechanisms produce the interference in Lindsay's study. But either account suggests that language understanding is slowed when people cannot engage in detailed embodied simulations, and, as a result, we can infer that these embodied simulations play some functional role in understanding. What functional role, the study doesn't tell us. All we know is that it's something that happens between recognizing the words and deciding whether they collectively make sense.

There have been other interference studies that point to the same conclusion. A study conducted by Michael Kaschak and his colleagues at Florida State took a slightly different tack.[2] They wanted to know not what would happen if you first saw motion and then had to respond to a sentence but what would happen when you had to do the two things at the same time. So they showed people moving images on a computer screen that went toward the participant, away from them, up, or down. And while the participants were watching these moving images, they had to listen to sentences that described motion in the same or a different direction. The moving images were created by having a spiral rotate clockwise or counterclockwise to create the illusion of forward or backward motion, and having horizontal lines move upward or downward to create the impression of downward or upward motion (static versions of these stimuli are shown in Figure 60). While each of these visual effects was on the screen, participants heard sentences and were asked to decide whether they made sense. These sentences could describe motion upward (*The cat climbed the tree*), downward (*The water dripped from the faucet*), toward (*The buffalo charged at you*), or away (*The car left you in the dust*). Other sentences, included to make the meaning-judgment task work, didn't make sense (like *The rib dropped a fire*). The expectation was that, like in Lindsay's study, it would take people longer to understand sen-

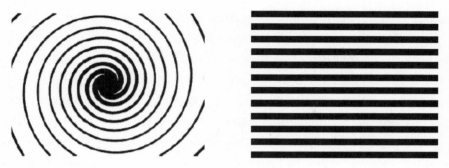

FIGURE 60 Static versions of the stimuli from Kaschak's study.

tences describing motion in the same direction as the motion being perceived on the screen. And indeed, although the effect was small (just 20 milliseconds), it was statistically significant.

Now, just like the Lindsay study, it's not clear exactly whether the interference effect results from adaptation of direction-specific cells or the perceptual incompatibility between the perceived event (e.g., a spiral moving away) and the described event (e.g., a car driving away). But it corroborates the findings that perceiving motion in a given direction interferes with understanding language about motion in the same direction, whether the sentence is presented simultaneously with the image or shortly after it.

One objection that a skeptic might raise to these last two studies is that they only show that people perform embodied simulation when they're asked to decide whether a sentence makes sense or not. Perhaps embodied simulation only plays a functional role in coming to a decision about whether a sentence makes sense but not actually in understanding it. In the real world, you normally don't need to make explicit judgments about meaningfulness. This objection basically says that asking people to decide whether sentences make sense focuses people's attention unduly to the meaningfulness of sentences, and perhaps it activates simulation as a by-product. In order to address this concern, Kaschak and his colleagues performed a second experiment, in which participants had to decide only whether the sentences were grammatical. That is, instead of making subtle meaning judgments, they only had to decide whether the sentence was well-formed Eng-lish or not. An example of an ungrammatical sentence might be some-thing like *The postman delivered a mail*. The nice thing about this different task is that it doesn't focus the participant's attention on the meaning of the sentence, because you can make simple grammaticality judgments like these without paying too much attention to meaning. If participants still took longer to respond to sentences describing motion in the same direction as the motion they were observing, even when not asked to make meaning judgments about the sentences, this would imply that embodied simulation plays a role in sentence understand-ing, even when people aren't driven by a task requiring special attention to meaning. And again, this is what the results showed—significantly slower reactions (by about 40 milliseconds) when the direction of the sentence and the visual illusion were the same.

Both of the studies we've just looked at show that people under-stand motion language more slowly when they're seeing motion or have just seen motion of a different object in the same direction (as do others).[3] These various behavioral interference studies all point to a common conclusion. If you occupy a flying pig's wings with swatting mosquitoes and that interferes with its ability to fly, then the wings probably play a functional role in flying. Likewise, impairing people's ability to construct embodied simulations interferes with their ability to understand language about visual or motor content, which implies that simulation does something. The current behavioral results don't show that embodied simulation is all that's needed for understanding (that it's sufficient for understanding) or that it's always required (that it's necessary). And they don't tell us what aspect of comprehension simulation contributes to. But they do show that it plays a functional part in the construction of meaning.

Physical Interference

If the behavioral interference studies discussed above are the equiv-alent of occupying a flying pig's wings so that it can't use them to their fullest while flying, then there's another approach, physical interference studies, which is roughly akin to tying the wings down or even lopping them off entirely. We know that simulation is executed, at least in large measure, in certain parts of the brain dedicated to controlling actions and in other parts dedicated to perceiving. And the brain is a delicate organ, which can suffer localized damage. What happens when these parts of the brain are temporarily or permanently unable to function? Does losing the ability to simulate performing an action hinder our ability to process language about that same action? Does losing the ability to perceive a particular type of spatial relation keep us from using language about that spatial relation? These are the questions that can be asked by studies in which embodied simulation is physically rendered impossible. The impairment can be produced by brain injury due to physical trauma like a car crash or health events like strokes. Or it can be induced temporarily in the laboratory.

Let's begin with language about objects located in space. You're now very familiar with how the visual system processes the locations of objects in space through the Where pathway. And given everything we've seen in this book, it should now come as no surprise that this part of the brain is also engaged when people use language about spatial relations. Using PET scans, Antonio Damasio and his colleagues at the University of Iowa scanned the brains of people who were looking at static pictures involving two objects in some spatial relation.[4] For instance, they might see a toaster on a table. They were asked to do one of two things—either name one of the objects (*toaster*) or name the relationship between the objects (*on*). What the PET scan showed was that the regions dedicated to perceiving spatial relationships in the Where pathway were active during this task and that this region was more active when they had to name the relation (*on*) than the object (*toaster*). In other words, the neural structures used for perceiving spatial relationships are also used when people produce language describing spatial relations. Pretty predictable so far.

But our question is whether this activation of the Where pathway serves a function when people are using language about space, and this study only tells us that the Where pathway is active, not whether it has a function. A series of studies by other researchers at Purdue University attacked this question by looking at people who had incurred brain damage that specifically affected their Where pathway.[5] They asked these people to perform a somewhat different task that also used prepositions like *on* and *in*. The participants were given sentences with one word missing and had to fill in the appropriate preposition. There were two types of sentences. One set of sentences used these prepositions spatially, for instance *John shopped for vegetables __ the store.* The other set of sentences prompted participants for the same preposition, but, in these sentences, the prepositions had meanings related to time, like *John shopped for vegetables __ nine o'clock.* Participants who had damage to the Where pathway performed more poorly on the spatial sentences than did people without brain damage. However, participants who had no damage to this region but did have other brain damage were able to competently produce prepositions for both sentence types. It seems clear from these studies that brain systems dedicated to perception play a functional role in language processing—losing the ability to use parts of the brain devoted to processing spatial relations impedes correctly using language about spatial relations but not necessarily other types of language. The Where pathway seems to be a functional part of using language about space.

The logic behind this finding is known as a *dissociation*, and it's the bread and butter of studies looking at the consequences of brain damage. It goes as follows. Suppose we observe damage to a particular brain region—like the Where pathway—and also observe selective impairment of a particular cognitive function A (spatial language use) but not another cognitive function B (time language use). We can then reason that this region plays a role in performing function A because when it is damaged, that function is impaired, but, when it is undamaged, that function works just fine. And we can also reason that the region in question is responsible for just this particular function because its impairment doesn't affect other cognitive functions, like B.

The dissociation is a leading tool for evaluating the relation between brain systems and their functions.

So if embodied simulation plays a functional role in language use, then what other sorts of dissociations should we expect to see? Let's start with a really simple case. Nouns, in general, are different from verbs in that nouns, like *pig, orangutan*, and *trucker hat*, often describe concrete, visible things, while verbs like *wiggle, stomp*, and *smile* tend to describe actions. So it's reasonable to suppose that different parts of the brain subserve accessing the meanings of nouns versus verbs; probably motor areas are more likely to be involved with verbs and parts of the What pathway are involved with nouns. And this is what we see in people whose brains are intact. Using PET, Harvard researchers took images of the brains of unimpaired language users while they performed simple tasks with concrete nouns or action verbs—producing singular or plural forms of nouns (e.g., *cat* or *cats*) or verbs (e.g., *run* or *runs*).[6] They found that using verbs selectively activated regions toward the front of the brain, in areas dedicated to motor control—the part indicated with the arrowhead on the leftmost image in Figure 61. But using nouns activated the What pathway, toward the lower part of the back of the temporal lobe, as seen on the right.

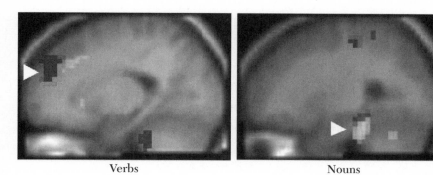

Verbs   Nouns

FIGURE 61

In other words, using action verbs activates motor areas, and using concrete nouns activates visual object identification areas. So far, so predictable. But to know whether these two brain regions play a functional role in using nouns versus verbs, we need to turn to dissociations. And there's good reason to believe that this might actually work, because it has long been known that, although some brain damaged patients will selectively lose their capacity to use nouns, others become poorer at using verbs. The remarkable—or perhaps at this point perfectly unremarkable—thing we find is that the brain lesions that lead to these deficits tend to be localized in the very places identified in Figure 61 by brain imaging in healthy people.[7] Patients who lose access to nouns are often the victims of damage to the left temporal cortex, the What pathway that represents objects, their shapes, and other properties. Patients who have difficulties with verbs tend to suffer from lesions in the left frontal cortex, an area dedicated to motor control. Not only do we have a dissociation here, we actually have a double dissociation. Damage to the frontal motor areas interferes with verbs but not nouns, while damage to the What pathway in the temporal lobe impairs nouns but not verbs. Case closed, right?

Maybe right. But maybe not. When people suffer brain damage, there's often an extensive region that's affected, and it could be the loss of any part of this region that is responsible for the language impairments. To really be sure that the part of the brain we're interested in is to blame, we need to make sure that only a particular piece of brain tissue is taken offline. And, for this, we need to turn from the study of naturally occurring brain damage to studies where the brain's normal functioning is impeded experimentally. We've already seen a method for doing this, transcranial magnetic stimulation, or TMS. (You'll recall that, in TMS, researchers apply a magnetic field through the skull to a specific part of the brain, which interferes very briefly with normal functioning of that brain region.) In essence, if applied correctly, TMS serves as a proxy for a very temporary brain lesion.

As of the writing of this book, no study has been done to our knowledge, using TMS to interfere with the What pathway in order to determine whether this selectively interferes with noun processing but not verb processing. However, there has been some work on

verbs. When you apply TMS to motor areas, lo and behold, performance on a task where people have to produce verbs is impaired but
producing nouns is not.[8] Seeing a dissociation like this due to magnetic stimulation of a specific brain region can only suggest that the
particular, targeted brain area is partially responsible for access to the
impaired type of language; in this case, motor areas play a functional
role in access to verbs.

So it appears that nouns rely on the What pathway and verbs rely
on motor areas. But it might have occurred to you that not all nouns
are about objects, just as not all verbs are about actions. *Marathon* is
a noun, even though it doesn't denote an object, and *imply* is a verb
that doesn't describe a physical action. So you might expect that different nouns, for example, would rely on different parts of the brain.
Ah, good, another prediction! When you look at different sorts of
nouns and who has trouble using them, you find that damage to different brain areas affects the use of different sorts of nouns in predictable ways. Nouns can denote all sorts of different kinds of objects
(among other things). Some of these objects we learn about primarily
through vision, like birds and faces. Others, like tools and toys, we
know about in large part through the motor interactions we've had
with them. We might therefore expect that brain damage to visual
versus motor centers would selectively impair access to words for objects that we know about mostly visually or motorically. Although this
precise idea hasn't exactly been explored, a closely related one has
been. One study reported on a double dissociation where words for
animals on the one hand and tools on the other were selectively impaired by damage to different parts of the What pathway.[9] It remains
to be seen whether damage to the motor system selectively impairs
access to nouns that describe things you know a lot about interacting
with physically.

Dissociations like these suggest that embodied simulation is not
just an irrelevant, downstream process. Embodied simulation plays a
functional role in understanding, at least for language about objects,
actions, and spatial relations.

Simulation Is Functional,
But Is It Necessary or Sufficient?

Here's what we know so far. People perform perceptual and motor simulation while they're processing language. They do so using the same parts of the brain they use to perceive the world and execute actions. Moreover, when specific aspects of embodied simulation are hindered, people have more trouble processing language about those specific aspects of perception or action. And finally, when brain regions dedicated to action or perception are damaged or temporarily taken offline, people have more trouble processing language about the specific perceptual or motor events it encodes. Taken together, all this evidence makes a pretty compelling case that embodied simulation plays a functional role in language understanding. The question remains open, though, whether embodied simulation is <u>necessary</u> for understanding, and whether it's all that's needed for understanding— whether it's <u>sufficient</u>.

Let's look at these issues in turn, beginning with the necessity question. Is it possible to understand language without performing embodied simulation? The answer might well be yes, in the right contexts. For instance, suppose you're hiking through a jungle, covered in low-hanging trees that are populated by notoriously aggressive monkeys. When your hiking companion yells *duck!* it's unlikely that you mentally simulate ducking and then reason about what your companion intended you to infer from this, and only then perform the appropriate action. No, instead of simulating, you probably go straight to <u>enacting</u>. Enacting is a very close cousin to simulating. As we've seen, hearing language about action engages the understander's motor system. So, given the right setting, the understander can simply stop suppressing actual action (unconsciously, presumably), so that instead of simulating the described action, the understander simply uses the motor system to perform the described action. In the *duck!* example, the understander might be prompted to allow enactment instead of simulation because of the context (heightened arousal due to the known presence of misanthropic monkeys) or by something in the way the command is yelled, like the volume, pitch, or rate. From an external perspective, we would say that a person who successfully and quickly ducks has understood the utterance, even though they might not have performed embodied simulation per se. So understanding need not involve embodied simulation if enactment takes priority over simulation.

There are plenty of other situations where it seems simulation might not be necessary. Consider, for instance, domains that you know really well. Suppose you're an engineer, and your drafter comes in and tells you that this morning he *drew up some schematics*. Do you really need to simulate what it would have been like for him to draw up schematics? If you've experienced this same scenario many times, then might you be able to simply jump to the conclusion that there are schematics that need looking at and that you should act or respond appropriately? Performing embodied simulation in this situation might not really help at all—if the inferences that you make when you encounter this language are predictable, then perhaps you will eventually learn to skip over embodied simulation, taking a shortcut so that the language you've processed directly activates the appropriate inferences. This might be faster and more efficient than performing an embodied simulation of the whole described scene.

So it seems that, even at first glance, there will be a variety of situations where embodied simulation might not be necessary for understanding—though these ideas still have to be tested empirically.

Let's turn to the question of sufficiency. Is simulation all you need in order to understand some piece of language? There's lots of language that seems to require more than just simulation. Consider for example language that doesn't describe scenes but instead serves to mark social interactions. For instance, how would simulation exhaustively explain what it is to understand the utterance *Hi!*? Even if there is some embodied simulation involved in understanding *Hi!*, there must be other things involved as well—reasoning about the type of social interaction that the person saying *Hi!* thinks they're engaged in, preparing for an appropriate response (like saying *Hi!* in response), or making inferences about the formality of the situation, the personal attributes of the speaker, and so on. Although some of this might involve simulation, it seems clear that understanding *Hi!* requires a lot of cognitive processing that embodied simulation of linguistic content couldn't provide. And this goes not just for *Hi!* but also for *Can I help you?*, *I now pronounce you man and wife*, *Like . . . whatever!*, and many other utterances that are closely bound together with specific types of social interaction.

Another type of language that might require more than embodied simulation is language about abstract concepts, as we discussed in the last chapter. Suppose you read a sentence like *Stock prices have dropped through the floor, and there are bargain basement deals out there for ambitious investors.* There's a lot of metaphor in here—stock prices can't literally drop though a literal floor, and, to my knowledge, there's no actual bargain basement where you can purchase stocks. As we saw previously, it's quite possible that embodied simulation plays a role—you might visually simulate something actually falling through a floor or a basement full of bargains. But this can't be the end of the story, because your understanding of the sentence goes well beyond the concrete images the metaphorical language might evoke. The inferences that you draw from language like this, which drive you to appropriate action, may be informed by embodied simulation, but they are informed by other processes as well. For instance, you need to know how to take inferences gleaned from one type of embodied simulation, say about changes in height, and apply them onto some other domain, like changes in value.[10] So in the case of abstract or metaphorical language, embodied simulation might play a role, but not a sufficient one.

If this is right—if embodied simulation plays a functional role in processing a lot of language a lot of the time but might not always be necessary or sufficient—then what should we think about embodied simulation in language understanding? Are we only interested in mechanisms that are necessary or sufficient? Should the fact that embodied simulation doesn't appear necessary or sufficient to comprehension lead us to dump it by the wayside? What types of questions should we be asking next?

It's instructive to compare language understanding with other similar cognitive functions that are better understood. Understanding language is a complex and heterogeneous behavior. So let's look at something similar but better understood, like vision. The human vision system is made up of a number of pieces, each of which plays an important functional role in allowing us to see. We need our eyes to collect and focus light on the retina, retinal cells to transform photons into electrochemical signals, axons to transmit these signals to the cerebral cortex, and the various parts of visual cortex to detect lines, colors, movement, objects, faces, and so on. Each of these parts of the whole is functional—if you lose the part of the vision system responsible for recognizing faces, for example, you can no longer recognize faces. But they're not sufficient. Every part plays a role, but any part by itself is a hopelessly useless chunk of brain tissue. In fact, it's hard to imagine a powerful cognitive function, like language understanding or vision, that could have any truly sufficient components. That's because these cognitive functions are, by their very nature, quite complex. They require many component processes housed throughout the brain to work in concert in order to function correctly. In this light, the insufficiency of simulation to language comprehension is quite unsurprising and shouldn't be much cause for concern.

On the other hand, the possible nonnecessity of simulation for language might be a very big deal. If we observe that some types of language don't require the use of embodied simulation in a functional way, this would suggest that understanding must be attainable through other means. Perhaps, to go back to the beginning of this chapter, there are Mentalese symbols floating around in people's heads, which come to life when simulation doesn't play a necessary role. How exactly they would make meaning is a big question and one that we'll have to leave open for now. But it raises the interesting idea that understanding with embodied simulation and understanding without embodied simulation might both be possible. And they might be different.

Functional, Sure. But What Function?

The analogy between language and vision is revealing. Trying to figure out whether the particular parts of the vision system play a functional role in vision is an important first step. But the really juicy part happens when we instead ask exactly what aspects of vision they functionally serve. The same is true for language. Once we've established that embodied simulation plays some role in making meaning, the next place our curiosity takes us it to what aspect of language, exactly, it contributes to.

This question is important for a variety of reasons. Some of them pertain to applications of the basic science. If we know exactly what certain types of embodied simulation do for exactly what aspects of language use, then we can better understand how it is that embodied simulation, or the lack thereof, contributes to the particular language impairments people with brain deficits display. We might even be able to develop therapies for people who have lost the ability to simulate in one modality or the other, so that these patients can develop work-arounds to regain language facilities that are more like normal. Likewise, if we want to build computer software capable of understand language like humans do, then knowing exactly how the different parts fit together will tell us how best to design artificial language users.

Identifying Words

In order to understand language, you need to figure out what words you're dealing with—you need to identify them. But words come at you really fast, and they are often less than perfectly intelligible, especially in speech, where environmental noises and speech errors frequently degrade the signal. But, as we've seen, over the course of a sentence, we simulate, early and often, which affects our expectations about upcoming words. So simulation might help ease the very demanding and sensitive task of identifying words by allowing us to predict which word is likely to come next. For instance, suppose you're talking in a noisy environment, say onboard an airplane, to the person seated next to you, and you can make out everything he says except the last word: *In my research with rabid monkeys, I've found that they're most likely to bite you when you're feeding them—you get little scars on your h___s.* Even if you failed to hear the last word, you could probably guess that it was *hands*. And that might be because you simulated how one might feed monkeys and what part of the body might be most accessible for them to bite.

Identifying Word Senses

Some words have multiple senses—for example, *ring* can refer to a boxing ring or a pinkie ring, just to name a few. Perhaps simulating along the way allows you to pick the right sense of a word with multiple senses on the first try. For example, if you read *The boxer slipped on his way into the ring*, you're probably not going to think that the right sense is the wedding one. Contrast that with the sentence *The boxer put on his coat and slipped on his ring*.

Representing Meaning

As we saw in the first chapter, most cognitive psychologists and linguists think that part of how we understand language is by activating internal, mental representations of what the language is about. But they differ in terms of how exactly they think those representations are encoded. Some have suggested that meaning is articulated in a language of thought, or Mentalese. Another possibility is that embodied simulations do the work of representing—they're internal reflections of purported external scenes. If this is true, then for those types of language that depend on simulation for mental representations, when simulation is impaired, people should fundamentally be unable to know and report what that language describes.

Creating the Subjective Experience of Understanding

We're not mere mechanical thinking machines. We also have subjective experiences—it feels like something to perceive a taste, to experience an emotion, or to understand language. Simulating might allow us to experience that internal feeling of knowing what something means. The basic idea is that we have subjective experiences when we're actually perceiving or actually moving our bodies. Reusing our brain systems for perception and motor control during language understanding may perhaps provide us with similar subjective experiences. What it feels like to understand language about a flying pig might be kind of what it's like to perceive a flying pig.

Drawing Inferences

Most of what we know from language is implicit, and we have to draw it out using inferences. Suppose your son comes home from school with a torn jacket and a bloody lip. His first words as he walks in the door: *You should see the other guy.* You probably infer that he had a fight, that he fared pretty well, comparatively, and that he's proud of it. None of that is explicitly mentioned in the sentence, but it could result from a simulation of what you think could have led him to look the way he does and say what he did.

Preparing to Act

Sometimes, it's really easy to know how to respond to language. Someone yells *duck!* and you do. But other times, you have to do some calculations to know exactly what the most apt response would be. Suppose you're helping a friend move, and she's got a particularly large, bulky, and unwieldy ottoman that you're carrying. She tells you *OK, this one goes in the guest bedroom, upstairs.* What's next? Do you try to carry it yourself? Do you ask for help—maybe this is a two-person job? You'll have to figure out whether you can do it yourself, whether it would be easier with two people, whether there's anyone available to help you, and so on. And maybe you simulate to figure out what the right thing to do is.

It's not surprising that language, as such a high-level, heterogeneous, and recently evolved cognitive function would piggyback off of other cognitive functions, like embodied simulation, or that these components would play functional roles. And, for that matter, it's not particularly surprising that these components might not be sufficient—just as specific parts of the vision are not sufficient to vision overall.

These are the big issues the field is dealing with today. What functions does simulation perform? And when does it perform these functions—are there some types of language that necessitate simulation and others that can make do just fine without simulation at all, thank you very much? This is the new science of meaning. It's a bunch of very smart scientists, with an array of tools at their disposal, trying to figure out the how and the why of our remarkable ability to make meaning.

CHAPTER 11

The Repurposed Mind

When you first opened this book (whether it was five hours ago or five years ago), you read this:

Polar bears have a taste for seal meat, and they like it fresh. So if you're a polar bear, you're going to have to figure out how to catch a seal. When hunting on land, the polar bear will often stalk its prey almost like a cat would, scooting along its belly to get right up close, and then pounce, claws first, jaws agape. The polar bear mostly blends in with its icy, snowy surroundings, so it's already at an advantage over the seal, which has a relatively poor sense of vision. But seals are quick. Sailors who encountered polar bears in the nineteenth century reported seeing polar bears do something quite clever to increase their chances of a hot meal. According to these early reports, as the bear sneaks up on its prey, it sometimes covers its muzzle with its paw, which allows it go more or less undetected. Apparently, the polar bear hides its nose.

I repeated this paragraph here to illustrate how far you've come over the last ten chapters. I'm not talking about how you're now aware that the polar bear's nose-covering behavior is most likely fictitious. No, I mean that you now know a lot about what you're doing when you read a passage like this. The sights and sounds, the actions, perhaps even the smells and tastes—you bring them to life through simulation. To do this, you pay close attention to the details of grammar, and you simulate early and often. These simulations are based specifically on your individual cognitive style and on your personal history of experiences, siphoned through your language and your culture.

The traditional theory of meaning—meaning as Mentalese symbols—doesn't really speak much to the two hundred–odd studies we've seen in these pages. It's not that anything we've seen definitely disproves the idea of a mental language of thought. That would be hard, because there's no obvious way to measure whether a Mentalese symbol is there. Instead, it's just that Mentalese doesn't buy us much. It doesn't predict that hockey players will use their premotor cortex to understand language about hockey actions. And it doesn't explain why Japanese speakers already have expectations about the shape of a mentioned object before they even come to the verb.

By contrast, the embodied simulation hypothesis does make very clear predictions about what we're doing when we're making meaning. It predicts that you should be doing all the stuff we now know you did when you read that paragraph about the polar bear's nose. And there's now some evidence that embodied simulations play a functional role in understanding—that we don't understand, or at least understand differently, when we don't simulate.

This isn't to say that meaning is conquered. We've only just arrived at the foot of the mountain. We still don't know exactly what embodied simulations are doing, functionally, for meaning. We still don't know exactly how meaning differs when simulation differs. And we still don't know whether we can make meaning without simulating. We don't know the answers, but we do know that these are good questions.

We know they're good because, first off, they're answerable. We can actually ascertain—and many of the studies covered in the preceding chapters have started to do this—what parts of the brain are being used for comprehension, under what conditions, and for what purpose. For instance, we know that the motor system is often used when people are understanding language about action, and that this is more likely when the language uses progressive than perfect aspect. We know that interfering with the perceptual system—by having people look at lines or spirals moving in one direction or another—affects how long it takes them to determine that a sentence makes sense, and so on.

We also know that these questions—about what brain mechanisms are used, when, and for what purpose—are good questions because they have clear real-world applications. When the brain is damaged, cognitive functioning can be impaired. Oftentimes, language is affected, and, because language is so important to everything we do from learning to socializing, it's really important to figure out how to help people recover their language abilities. The better we understand exactly what certain parts of the brain do for language use, the better we'll be able to diagnose brain injury from their linguistic impairments and to develop treatment and recovery strategies for patients. There is also a range of developmental disorders that affect language, including specific language impairment and autism spectrum disorders, where again diagnosis and treatment would benefit from a better understanding of what parts of the brain do what when for language comprehension and production. And finally, knowing more about how the brain makes meaning is important if we want to make machines smart like us. If you're trying to build a piece of software that not only behaves like a human but also understands language like a human, you'll need to build in something like the solutions that humans use to make meaning.

As you might have noticed, these productive questions happen to be the ones that this book has been asking throughout. (Pat on the back for the author.) The field has many careers of work ahead of it to answer them thoroughly, but early glimpses of answers are already taking form. Dawn has broken on a new era in the science of the mind.

The Age of Meaning

In many ways, in the cognitive sciences, the twentieth century was the age of form. It was the century in which people invented ways to automate computational operations, first in hardware, then using software. Soon thereafter, we came to apply the conceptual framework of the computing machine to the study of the brain and the mind, viewing it as performing formal syntactic operations, although in a massively parallel and probabilistic way. And this included the study of how the brain computes language. But there was also a second, more language-relevant way in which the twentieth century was the century of form. Certain aspects of language (and not others) are more amenable to study from the perspective of the mind as a computer. As a result, those aspects of language came to be studied the most. These include the form of words (their sound structure, or *phonology*, and their compositional structure, or *morphology*), and the form of sentences (*syntax*).

But this is a new millennium, and the twenty-first century has ushered in renewed attention on not just form but what the form does—now that we've mastered a lot of how formal operations work, we're able to focus our attention on what they mean.[1] In the study of language and the mind, this translates into not merely studying the form of language (though we still care a lot about how grammar works) but also its meaning.

This turn to meaning isn't easy, because meaning is still hidden, variable, and hard to categorize. But it's inevitable. Conveying meaning is what language is for. It's the reason the capacities we have to use language evolved in our species; it's the reason children acquire it; it's the reason we ever bother to utter or listen to a word. To study language without meaning would be like studying running without motion, eating without hunger. How could we afford not to study meaning, when it is one of the ways that human language is unique among nature's communication systems? How could we avoid it when it is so intensely personal—when it tells us so much about who we are? Meaning is what our personal experiences of language are about and what makes them unique to us as individuals. The turn to meaning is inevitable because meaning is what makes us uniquely, cognitively, human.

This isn't to say, of course, that the study of the form of language is going away. But in a science of language where meaning has a more central place, the study of form will change—it will bend around the gravitational tug that meaning exerts. That's because there's a close relation between the form of language and its meaning. People don't randomly choose to use an active form of a sentence or a passive one—they often make those choices for meaningful reasons. People don't utter sentences in just any sequence; order matters to meaning because people process a sentence differently if you say *I ate my lunch after reading the paper* than *After reading the paper, I ate my lunch*. Linguistic form is affected by meaning. And as a result, a science of language that focuses on meaning won't obviate the study of form; it will make it more explanatory. It will explain more of the existing data that we couldn't understand without meaning (why do people sometimes use active and sometimes passive?) and make better predictions about future observed data. In other words, more focus on meaning will lead to a better science of language.

By all present appearances, the twenty-first century is off to a good start as the age of meaning. And this can't help but be a good thing. It allows us to study language in its complete and natural state as one particular, extraordinary way that our species manages to communicate.

How We Got to Be This Way

In most ways, the human species is thoroughly unremarkable in the tree of life. We can't outrun most large mammals, we don't have particularly interesting plumage, and our fangs and claws fail to strike fear in the hearts of large predators, or for that matter even our domestic pets. And yet, among all the animals that have yet graced the planet, we do have something special. We're smart. And not just chim-panzee-figure-out-how-to-jab-termites-out-of-a-hole-with-a-stick smart. We're build-a-rocket-ship-to-send-a-chimpanzee-into-space smart. And a large part of that smartness has to do with language.

We alone have the capacity for human-scale language. Other ani-mals can communicate—bees can waggle messages to each other about where the good flowers are, and dolphins, well, who knows what dolphins are doing, but it sure seems like they're communicat-ing. But take any nonhuman animal and try to teach it any form of human language—spoken, written, signed—and the animal is going to fall seriously short. Some chimps and dogs can learn as many as hun-dreds or thousands of words, which is pretty remarkable, and maybe even some rudimentary syntax.[2] But none of them are going to be able to figure out that it was actually the lawyer being cross-examined or that stock prices aren't literally falling. And that's true even if you raise a chimp exactly as a human child. We know this from the experience of the Kelloggs, two scientists who conducted that experiment—a real-life *Bedtime for Bonzo*—in the 1930s.[3] Humans are biologically different from other animals in a way that allows us, and not them, to do ridiculously hard things with language, like read this paragraph. And that biological difference is the product of evolution.

How did we get this way? What happened over the course of the evolution of the species that took an animal that was also the precur-sor of chimps and bonobos and shaped it into us?[4] What is our unique genetic endowment? Did a random mutation cause early hominids to sprout a nascent mental language organ—an independent new growth simply added to the existing primate organism that we were? Or were the existing parts of our primate brains tweaked in subtle ways that multiplied the uses we could put them to?

Sadly, we don't know for sure. But at least with respect to meaning, the new science I've described in this book does give us some clues. Meaning is subserved by a variety of cognitive systems that predate language. To convey thoughts and intentions and ideas, we perform embodied simulations recruiting brain areas that, among other things, we also use when perceiving or acting. These are in fact brain systems that have very close homologues in other animals. We make meaning using these evolutionarily old parts of the brain that, over the course of hundreds of millions of years of mammalian evolution, came to be specialized for functions other than language—functions like perceiving and acting.

In other words, the systems we use for meaning don't seem to have spontaneously sprouted ex nihilo to subserve only language. Instead, in the human capacity for understanding language, evolution has cobbled together a new machine from old parts it had lying around in the junkyard of the brain. As far as meaning goes, we are distinguished from other animals not in that we've evolved a brand new mental organ but rather in that we have recycled older systems for a new purpose. We use our primate perception and action systems not only when we're actually perceiving or acting but also when we're understanding language about perceiving or acting. Evolution has repurposed parts of our brains.

This cuts to the core of how we came to be the way we are. We aren't merely smart chimpanzees who evolved new brain systems for meaning. We're smart chimpanzees who evolved the ability to use the brain systems we already had in new ways. And this fits in well with what we know about how evolution works more generally. Evolution is a tinkerer. Using existing systems in new ways, that is, exapting them, is a pretty good way to quickly and incrementally develop cognitive machinery that can perform a complex function, like understanding language. And it seems to be what evolution has done.

But it's not as simple as that. Think about what it would take to augment a motor control system so that it could also be used for understanding language about motor control or, for that matter, for imagining or recalling motor control. Clearly, the circuitry would have to be modified in some way, because otherwise every time you thought about moving, you'd actually move. And the same goes for perceptual simulation; you wouldn't want to hallucinate a polar bear in front of you every time you thought about one—it's hard to imagine that

conferring an advantage in propagating genes. No, the motor and perceptual circuitry had to be tweaked a little bit so that it could be run
"offline," as it were, so that sometimes you can activate premotor cortex and yet not send a signal to your muscles to contract.

There are several different ways that this could have come about,
and we yet don't have evidence to adjudicate among them. One possibility is that simulation could work similarly to something that happens while people are asleep. During REM sleep, people have vivid
dreams, many of which are about moving their bodies, and, like imagery, memory, and language use, this is achieved with activation of
perceptual and motor systems of the brain. But, except in the case of
certain pathologies, the muscles of REM sleepers don't actually act
out the dream actions. This is due to something called REM atonia,
the phenomenon where motor neurons, the neurons that actually control the firing of muscles, do not themselves fire. There's still substantial debate about the exact biochemical causes for REM atonia.[5]
But, whatever they are, they (or some other mechanisms like them)
could ultimately also be responsible for our ability to use motor and
perceptual systems for language without acting out the simulated actions or hallucinating the simulated images. The basic idea is that, by
inhibiting the firing of motor neurons, which are the last way station
that conveys a volley of neural activation to muscles, we could use the
motor system any way we wanted, without having the repercussions of
downstream action. This would allow simulation to happen offline,
for tasks where we wanted to think about moving but not actually
move. And the same would go for perception.

Ultimately, we don't yet know exactly how evolution has changed
the human brain to afford the extensive and pervasive simulation it
appears to perform. Perhaps the changes already started showing up
in a much more limited way in human ancestors, such that they're
now reflected in a more limited way in nonhuman primates.[6] Or perhaps extensive use of embodied simulation is a capacity that coevolved
with language, thus specifically a recent development along the ho-
minid line. But we do know that our use of simulation is extensive,

throughout language and other higher cognitive functions that we uniquely display among animal species. In that way, simulation—or perhaps merely the pervasive extent to which we use it—makes us uniquely human.

Beyond Language

Humans haven't repurposed motor and perceptual systems for language alone. We saw earlier how simulation is vital to a variety of the cognitive functions that we think of as part of the quiver of faculties that make us human, like memory and imagination. But that's only the beginning. Even the most far-fetched, least likely aspects of human cognition seem to be shaped by simulation. Take math. We like to think of math as one of the pinnacle demonstrations of abstract complex cognition; Plato even said that it was the "highest form of pure thought." But as you might now be able to guess, there is substantial evidence in recent years that math, like language, recruits evolutionarily older systems, like those tuned for perception and action. It seems that when people are actually doing math, they use brain regions that also do spatial processing.[7]

There are different types of evidence. One of the most frequently cited is known as the SNARC, or "spatial-numerical association of response codes."[8] Don't worry too much about the clunky name. This is just the finding that people are faster to respond to larger numbers on the right side of their body and to smaller numbers on the left side. This is true whether the number is presented visually in front of them to one side or the other, or whether it's presented in the middle of their visual field and they have to respond to it by pressing a button with either their right or left hand. This association of space with number has led some researchers to conclude that people think about number using a mental number line. And there's more convergent evidence. When you have people pick a number (any number), they tend to look left when picking lower numbers and right then picking higher numbers.[9] And so on.

Even more compelling work comes from patients with hemispatial neglect, which is the phenomenon where people suffer damage to one hemisphere of the brain, impairing their ability to attend to one side of space around them. People with hemispatial neglect have trouble perceiving and processing stimuli on one side of their body, so they might eat only the food on the nonneglected half of their plate or, when asked to bisect a line depicted in front of them, will tend to indicate the midpoint as being farther to the nonneglected side. So we might ask, in the spirit of dissociation studies, whether hemispatial neglect also affects people's reasoning about numbers. If so, then this would mean that cognitive systems that do spatial cognition are functionally involved in math. Amazingly, people with hemispheric neglect also seem to have a sort of numerical neglect.[10] If you ask people with hemispatial neglect to say what number is exactly between 2 and 6, they will tend to shift their answer in the "direction" of their nonneglected side; so someone with left hemispatial neglect (which is more common than right neglect) will be more likely to answer 5 than the expected 4. People seem to use some of the same brain circuitry that allows them to perceive and act in space in order to make judgments about numbers.

And as we run down the list of higher human cognitive functions, like reasoning, problem solving, decision making, and so on, we may well find that these work the same way, that these functions are cobbled together at least in part by using relatively older, lower-level systems in new ways. When people make judgments about how similar two abstract concepts are (like *commitment* and *justice*), they think they're more similar the closer together in space they're presented.[11] When people make judgments about how important a political leader is while simultaneously holding a physical object, the weight of the object affects the importance they ascribe to the leader.[12]

Perhaps using systems for perception and action is even a hallmark feature of human cognition. Maybe as we've gotten smarter, we've gotten better at turning our finely-tuned primate perception and action systems to other purposes, of which language is only one particularly salient example.

What Is It to Communicate?

The new science of meaning asks us to reconsider a lot of the truths about language and the mind that we held to be self-evident. And one of these is the fundamental question of what communication is. The lay notion of how communication works—which is also the basis for the language of thought hypothesis—is basically that we have ideas in our heads that we translate into language and then transmit to others using that language, such that they can extract the information and contain it in their own heads. There are a bunch of assumptions at play here. First, the information the speaker wants to encode is discrete and delineable. Second, there's a best set of linguistic choices that a speaker can make to uniquely identify the information she wants to convey. And finally, if the hearer activates the right decoding processes, she can transform those words into the right information. Communication succeeds!

There may be some merit to this view, but the evidence on embodied simulation muddies the waters significantly. As speakers, the messages we intend to transmit are probably far from discrete packets of information. Instead, they are dynamic and continuous currents of perception and action, either performed and perceived or mentally simulated. As speakers, we have to jam all this messy, amorphous, nonspecific, continuous stuff through the narrow aperture afforded by discrete words and grammatical structures available to us in our language. This process of encoding is necessarily lossy: a few words can't hope to capture the breadth and depth of the perceptual, motor, or affective experiences we want to convey. It's also nondeterministic: we might use two different words to describe the same thought, sometimes even in the same sentence. And it's fickle: we often just reuse words or grammar that we've just uttered and we tend to mimic the linguistic patterns of our interlocutor, instead of picking out the theoretically perfect words for a given message. So the information we want to convey is neither neatly delineated to begin with nor uniquely and perfectly packageable in words.

On the part of the hearer, we know that at least part of what people do to decode an utterance is to build an embodied simulation. But this embodied simulation is radically underconstrained by the words and grammar present in the utterance. We have to make a lot of assumptions and dig deep into our world knowledge in order to come up with a way to understand even something as concise and simple as *The polar bear hides its nose.*

We've also seen that people differ dramatically in the way they bring different resources to bear on comprehension, based on individual differences in cognitive style or different experiences due to intra- or cross-cultural variation in the way they have used their bodies in the world or the things they're likely to have perceived. Not to mention the likely possibility that people will have learned different associations from words to simulations. Concretely, this means that even if we pick words that—for us—uniquely identify a particular simulation, there's no guarantee that the hearer will do the intended thing with those words in decoding.

For example, while writing this book, I became aware that somehow, over the course of my not insignificant exposure to the English language, I had come to believe that the expression *cut from whole cloth* meant something like "built from scratch." So in my idiolect (my own version of English), something *cut from whole cloth* would contrast with things (like the human capacity for language!) that are built from preexisting parts. But apparently, for some of the English-speaking world, *cut from whole cloth* means "made up, as a lie." Others I've subsequently asked have reported that they've never heard it before.

So how on earth are we ever able to communicate when we don't even agree on what the words we use mean, and, when we do, we may or may not engage the same cognitive processes in order to flesh them out? It seems as though it's vanishingly improbable that a speaker, intending to convey some message (under the pressure of the time needed to plan and articulate an utterance before losing the hearer's attention and trying their patience), will pick just the right words that evoke just the intended simulations in the hearer's mind.

And yet somehow we do communicate. So something about the model must be wrong. And when you think about how people use language—what communication is actually for and how it's situated inside human interactions more broadly—an explanation starts to present itself.

What is it to communicate successfully? Clearly, because we can't see into others' minds, the metrics we use have to be indirect. Did the hearer do what I wanted him to? Did he act as though he understood? Did he respond verbally in a way that implies that he understood what I intended? The key is that for these types of metrics, which are what we mean, practically, by "successful communication," it's not necessary that the message in the speaker's head be exactly replicated in the head of the hearer. (Which is good, because there's no way for us to tell if it were.) Instead, all that's needed is a sufficient degree of commensurability between speaker's intent and hearer's interpretation. If the words I select as the speaker cause you to engage some cognitive processes (whatever they might be) that lead you to act behaviorally as though you've understood, then it doesn't matter how similar what I did before or during encoding is to what you did during or after decoding. The two processes need to be functionally equivalent to a sufficient degree, but there's no reason that they have to be mirror images of one another.

This would start to explain why people can conjure up different simulations to understand similar language, without communication breaking down completely. When we probe deeply, we might find behavioral indicators that a visualizer and a verbalizer understand an utterance differently, but, most of the time, for most of the purposes we put language to, we don't see any external signs of these differences, and so it appears not to matter. To successfully communicate, you don't have to get it just right, you just have to get it close enough.

Sometimes, of course, the system will break down. You think of *waiting on a bench* as sitting, while I think of it as squatting. And, as a result, when I tell you that I got bubble gum on my backside while I was waiting for you on the corner, you assume I must have sat in it, but in fact I intended you to infer that someone must have come by and vandalized me midsquat. And failures to communicate are certainly common. People with different degrees or types of experience with domains under discussion will have incommensurate simulations of described events, which may lead to difficulty in communication.

But this, like most of the open avenues discussed in this last chapter, remains to be fleshed out experimentally. The good news is that this is the century in which cognitive science is really sinking its teeth into meaning. And it's a topic that, like a hard-earned meal of seal meat, offers no lack of sustenance.

The Crosstalk Hypothesis

For almost as long as there have been telephones and automobiles, people have been trying to put one into the other. There's now overwhelming evidence that this is not a good idea. Driving a car while you're talking on the phone makes you do pretty much everything worse in the driver's seat, from braking on time[1] to steering successfully.[2] You're even a worse conversationalist![3] And although texting and other activities that take your attention off the road are the worst culprits, you're also a bad driver even when simply talking on a hands-free device.[4]

You've probably seen characteristic cell-phone driving behavior; it's kind of like a sloth on Quaaludes who's occasionally getting randomly applied electroshock therapy. You'll see someone pull up to a stop sign, coming to a complete stop. He has the right of way, but inexplicably he just sits there. And sits. And you're thinking, *Do I pull around him? Did he fall asleep?* And then all of a sudden, he starts drifting into the intersection. At this point, there's another car coming from the other direction, so he has to slam on his breaks. You're watching all this, thinking, *What is this guy on?* But you know exactly what's he's on. A cell phone.

Legislators are starting to get the message. As of late 2011, thirty-five states have banned at least some type of mobile device use while driving. And, following a series of high-profile fatal collisions between cars and trains and other large objects, the National Transportation Safety Board has now recommended that states ban all use of portable devices—that includes everything from texting to talking on a hands-free device—except in the case of emergencies.

So we know that it's hard to drive while talking on a cell phone. But the question is: Why? What is it about talking on the phone that interferes with seeing the world and moving your body?

You might think that this isn't a very interesting question—after all, there are lots of things you can't do well at the same time. You probably can't pat your head while rubbing your belly. You most likely can't remember the lyrics to "Three Blind Mice" while singing "The Star Spangled Banner." You probably can't remember a new phone number you just heard while counting backward from a hundred.

But notice that these three pairs of incompatible things all share one property. In all three, the things you're trying to do at the same time have something in common. Patting your head and rubbing your belly are both arm movements. Singing and remembering lyrics both require you to recall words in a particular sequence. And counting and remembering numbers both involve thinking about numbers. The reason that these different activities interfere with each other is that they ask a particular part of your brain (the part that controls arm movements or recalls words or represents numbers) to do two different things at the same time. The parts of your brain are like the other parts of your body. You can't use your mouth to chew and whistle at the same time, because the two actions are physically incompatible. In the same way, you can't use the same part of your brain to run two different operations at the same time.

On the other hand, there are some things that are easy to do at the same time. You can walk just fine while chewing gum, you have no problem doing push-ups while listing all the states you've visited (within the limits of your upper-body strength and long-term memory), and it's hardly a hindrance to scratch your back while reciting the Pledge of Allegiance. What makes these things easy to do together is that they don't use the same parts of the brain.

So now we come back to talking on a cell phone while driving. Driving experts have been baffled for a long time about why it is that talking and driving are hard to do at the same time. After all, to use a cell phone, what do you do? You talk using your mouth and listen using your ears. To drive, you use your eyes and hands and feet. There's isn't any obvious overlap there. It seems like there should be no interference. And yet there is. What's going on?

It should be no surprise what I'm going to say next. I think that embodied simulation accounts for at least part of the interference you see when people are driving and talking. After all, any language—literal or metaphorical—that engages motor and perceptual simulations has the potential of interfering with the visual perception you need to see the road, the use of your hands you need to steer the vehicle, and the use of your feet required to control your speed. Of course, there are probably other reasons why it's hard to talk and drive. Maybe part of it is just the global attention you need to do the two tasks at the same time. Or maybe it's that you sometimes even prioritize the conversation over the driving (especially if you're a teenager). But there are now a couple of studies out showing that not all language is equivalent in how it affects driving, and that visual and spatial language in particular interfere with driving more than other types of language.[5] That could mean that embodied simulation is part of what makes it so hard to drive and talk at the same time.

My point in bringing this example up is just that embodied simulation isn't merely something limited to the tightly constrained conditions of a controlled laboratory experiment. Making meaning is something we do throughout our daily lives. And sometimes, if we're doing it in the wrong way, at the wrong time, it can have powerful consequences. I'm hopeful that as the new science of meaning matures, we'll see more research into its various applications, from the driver's seat to the cockpit to the classroom and beyond.

ACKNOWLEDGMENTS

If there's one person most to blame for the book you have in your hands (said the author, passing the buck), it's George Lakoff. I met George when I was a seventeen-year-old undeclared, overcaffeinated undergrad. He mentored me through three degrees. It's no exaggeration to say that he not only laid the intellectual groundwork for almost everything I have to say in this book but that he's also the reason I work in the field at all. The example of George's tireless work ethic, commitment to big ideas, ingenuity in solving unsolvable problems, and dedication to being above all things a thoughtful, kind, decent human being—a mensch—have made me a better scholar and a better person. To the extent that there is anything valuable in these pages, it's in large part due to George. That isn't to say that I expect George to agree with everything I have to say in this book. On the contrary—he will probably object vehemently to some of it. If there's anything I've learned from George, it's that disagreement is the beginning of learning. I hope this book serves as the basis for many fruitful arguments.

I've also been fortunate to have had other intellectual mentors who have contributed in no small way to how I think about language. Jerry Feldman is the model of a big-picture thinker who at the same time has the technical acumen for the details and cares about the relation between the two. He's helped me think about computational and implementational issues for many years. John Ohala was another of my early mentors. This book might not be entirely up his alley, since his research is on the other end of language (how we perceive and produce sounds). But I learned the value and practices of laboratory research under his guidance and can only hope that the work I'm now doing approaches his exacting standards. Dan Slobin has also guided my interests over the years, showing by example how attention to linguistic detail can reveal linguistic and cognitive diversity and systematicity.

There are others people I'd like to thank who have guided my thinking by being leaders in this emerging field, and I would have very little to say here without their brilliance—I would also be less able to say it clearly in some cases if it weren't for their generosity with feedback. Larry Barsalou, Rolf Zwaan, Art Glenberg, Stephen Kosselyn, Ray Gibbs, Lera Boroditsky, and Friedemann Pulvermüller are inspiring figures whose shoes I wouldn't dare try to fill but on whose shoulders I'm more than happy to stand.

Science is a collaborative process, and most of the research I describe in this book was done in teams with colleagues. For their industriousness, patience, and insight, I thank Nancy Chang, Teenie Matlock, Amy Schafer, Rafael Núñez, Seana Coulson, Srini Narayanan, Shane Lindsay, Nate Medeiros-Ward, Dave Strayer, and Frank Drews. It's much easier to do good science when you have teammates who compensate for your own inadequacies.

A number of students have helped shape my thinking by challenging my ideas, bringing in their own perspectives, and, concretely, reading drafts of chapters. I'd like to thank Heeyeon Dennison, Manami Sato, Ting Ting Chan Lau, Kathryn Wheeler, Nian Liu, Jinsun Choe, Bodo Winter, Meylysa Tseng, Wenwei Han, Tyler Marghetis, Ross Metusalem, and Tristin Davenport. Clarice Robenalt was immensely helpful in tracking down references and constructing the index. She probably knows what's in this book better than I do at this point.

Many of the early chapters of this manuscript were tempered into shape in the crucible of a writing group. To aspiring writers of anything—from poetry to scientific articles to books about language—I can't recommend anything more strongly than conferencing your writing. My writing partners Ashley Maynard, Lori Yancura, and Katherine Irwin shaped the style and presentation of the core of the book. They provided a constant reminder that not everyone is an expert on this stuff, and their suggestions about how to reach a nonexpert audience were invaluable. We also had a lot of fun.

Later in the writing process, several friends and colleagues graciously gave me feedback on drafts, and I'm indebted to Adam Ruderman, Madelaine Plauché, Richard Carlson, Marta Kutas, Sian Beilock, Ray Gibbs, and David Kemmerer. Any mistakes of course are my own doing, but at least there are fewer of them in these pages, thanks to these generous people.

Many readers will wonder how this book ever got published, not least myself. In large part, the blame falls on Katinka Matson and the staff at Brockman, who took a chance on a naïve and shoddily constructed book proposal and put in many, many painstaking hours helping me pound it into coherence.

That brings me to T. J. Kelleher and Tisse Takagi at Basic Books, who provided invaluable feedback at the high level and low. They helped me formulate my ideas into a coherent and articulate manuscript. I owe them a debt of gratitude, as I do to Collin Tracy for shepherding the manuscript through production and Carrie Watterson for her careful copy editing.

Finally, of course, there's my family. Writing a book is the most painful thing I've ever done, and that's saying a lot, because I've done a lot of painful things, including once breaking an elbow while snowboarding and another time watching the entire fourth season of *Battlestar Gallactica* in one sitting. I couldn't have done it without the love and support of my parents, grandparents, brother, siblings-in-law, and most importantly my partner, Frances. Thank you for keeping me tethered to the material world while I've been so deeply immersed in the world of ideas.

NOTES

Notes to Chapter 1

1. Spectator, 1901, p. 984; Lopez, 1986
2. OK, I'll admit it. I didn't read about it first, I actually heard about it first in Matt Damon's internal monologue in The Informant! THEN I read about it.
3. There are even technical terms for the two parts—*intension* versus *extension* or *signified* versus *reference*.
4. Glenberg, & Robertson, 2000
5. This is a very old idea. It's most recently been associated with work by Fodor (1975) and Pinker (1994), but goes back to Bertrand Russell's work on logical systems (1903).
6. Searle, 1999
7. As well as Ron Langacker, Giles Fauconnier, and Adele Goldberg.
8. Like Merleau-Ponty, Andy Clark, and Mark Turner.
9. Like Francisco Varela and Arthur Glenberg.
10. Larry Barsalou told certainly the most influential version of the story in his 1999 paper in *Behavioral and Brain Sciences*.
11. The discovery of so-called *mirror neurons* (which we'll look at in Chapter 4) led Gallese, Rizzolatti, and their colleagues in Parma to speculate about how meaning works in this paper: Gallese, Fadiga, Fogassi, & Rizzolatti (1996).
12. Two dissertations that appeared in 1997 by David Bailey and Srini Narayanan led the way here.
13. Stirling, 1974

Notes to Chapter 2

1. Hartgens & Kuipers, 2004
2. Driskell, Copper, & Moran,1994; Weinberg, 2008
3. Woolfolk, Parrish, & Murphy, 1985
4. Perky, 1910
5. Farah, 1985
6. Craver-Lemley & Arterberry, 2001
7. Farah, 1989
8. Segal, 1972
9. Finke, 1989
10. Shepard & Metzler, 1971

11. The two objects in both A and B are the same; the objects in C are mirror images of each other.
12. Cooper & Shepard, 1973
13. Cooper, 1975
14. Kosselyn, Ball, & Reiser, 1978
15. Olin, 1910
16. Penfield, 1958
17. Kraemer, Macrae, Green, & Kelley, 2005
18. Halpern & Zatorre, 1999; Wheeler, Petersen, & Buckner, 2000
19. Porro et al., 1996
20. Ehrsson, Geyer, & Naito, 2003
21. Lotze et al., 1999
22. Roediger, 1980
23. Maguire, Valentine, Wilding, & Kapur, 2003
24. Wheeler et al., 2000
25. Nyberg et al., 2001
26. Solomon & Barsalou, 2004
27. Pecher, Zeelenberg, & Barsalou, 2003

Notes to Chapter 3

1. In this book, I'm mostly focusing on how people make meaning in language comprehension, because there has been substantially more research on meaning in comprehension than production. The reason for this is that, for the most part, it's easier, using the tools we have at our disposal, and also more revealing of the underlying cognitive processes, to present language to people and measure what happens than to try to make people think certain things and measure what they say.
2. Lakoff, 2004
3. Stanfield & Zwaan, 2001
4. Zwaan, Stanfield, and Yaxley, 2002
5. Ibid.
6. Wilcox, 1999
7. Vandenbeld & Rensink, 2003
8. Aginsky & Tarr, 2000
9. Connell, 2007
10. Bergen, Lindsay, Matlock, & Narayanan, 2007
11. Ibid.
12. Here's how it works. First, a light (usually infrared) is pointed at a person's eye, and the light that reflects off the eye is collected by a high-speed video camera. The movements of the eye are then calculated by changes of certain points on the eye. Infrared light makes the pupil appear white, and the pupil, naturally, moves as the person looks around. However, the infrared light also leaves a direct reflection on the cornea of the eye—the transparent covering around the eyeball. This corneal reflection does not move when the eyeball moves, so it serves as a landmark for calculating where the pupil is oriented relative to the head. Other equipment (often using magnets placed on the headpiece) detects where the person's head is oriented relative to objects of interest in the world. Software then combines the representation of where the head is located and oriented and where the eye is pointing relative to the

head to determine where the person's gaze is pointing, to produce an integrated recording of where the eye looked when over a field of vision.

13. Spivey & Geng, 2001
14. Although it's unethical to lie to participants in your experiments if there's any chance they could be harmed, many experiments in which there's minimal risk to participants ethically invent harmless and elegant deceptions like this one.
15. Johansson, Holsanova, & Holmqvist, 2006
16. Richardson, Spivey, Barsalou, & McCrae, 2003
17. Zwaan, Madden, Yaxley, & Aveyard, 2004
18. Burgund, & Marsolek, 2000
19. Bergen & Chang, 2005; Feldman, 2006; Lakoff, 1987; Zwaan, 2004; Zwaan & Madden, 2005
20. Palmer, Rosch, & Chase, 1981
21. Blanz, Tarr, & Bülthoff, 1999
22. Edelman & Bülthoff, 1992; Tarr, 1995; Tarr & Pinker, 1989
23. Cutzu & Edelman, 1994
24. Borghi, Glenberg, & Kaschak, 2004
25. Yaxley & Zwaan, 2007
26. Horton & Rapp, 2003
27. Winter & Bergen, 2012
28. Zwaan, 2004, p. 38

Notes to Chapter 4

1. Ramachandran, 2006
2. Gallese, Fadiga, Fogassi, & Rizzolatti, 1996; Rizzolatti, Fadiga, Gallese, & Fogassi, 1996
3. Iacoboni et al., 1999
4. Bergen & Wheeler, 2005; Borregine & Kaschak, 2006, Glenberg & Kaschak, 2002; Glenberg et al., 2008; Kaschak & Borregine, 2008; Tseng & Bergen, 2005; Zwaan & Taylor, 2006
5. Glenberg & Kaschak, 2002
6. Bergen & Wheeler, 2005
7. Bergen & Wheeler, 2005
8. Masson, Bub, & Warren , 2008
9. Siakaluk, Pexman, Aguilera, Owen, & Sears, 2008
10. Zwaan & Taylor, 2006
11. Buccino et al., 2005
12. Pulvermüller, Haerle, & Hummel, 2001
13. Hauk, Johnsrude, & Pulvermüller, 2004
14. Tettamanti et al., 2005

Notes to Chapter 5

1. Treisman, 1996
2. Hagoort, 2005
3. Pfanz, Aschan, Langenfeld-Heyser, Wittmann, & Loose, 2002
4. Lakoff, 1987; Langacker, 1987; Talmy, 2000
5. Goldberg, 1995

6. Kaschak & Glenberg, 2000
7. Goldberg, 1995
8. Chomsky, 1955
9. Arnold, Wasow, Losongco, & Ginstrom, 2000
10. Nigro & Neisser, 1983
11. Libby & Eibach, 2002, Libby & Eibach, 2007
12. Ruby & Decety, 2001; Vogeley & Fink, 2003
13. Zwaan et al., 2004
14. Chan & Bergen, 2005; Chatterjee, 2001; Chatterjee, Southwood, & Basilico, 1999; Maass & Russo, 2003
15. Brunye, Ditman, Mahoney, Augustyn, & Taylor, 2009
16. Comrie, 1976; Dowty, 1977
17. Magliano & Schleich, 2000
18. Madden & Zwaan, 2003
19. Magliano & Schleich, 2000
20. Carreiras, Carriedo, Alonso, & Fernández, 1997
21. Madden & Zwaan, 2003
22. Strohecker, 2000
23. Borghi et al., 2004; Denis & Kosslyn, 1999; Mellet et al., 2002
24. Bergen & Chang, 2005; Chang, Gildea, & Narayanan, 1998; Madden & Zwaan, 2003
25. Bergen & Wheeler, 2010
26. Bergen & Wheeler, 2005; Borreggine & Kaschak, 2006; Glenberg & Kaschak, 2002; Tseng & Bergen, 2005
27. Svensson, 1998
28. Boroditsky, Schmidt, & Phillips, 2003

Notes to Chapter 6

1. Trueswell, Tanenhaus, & Garnsey, 1994
2. Altmann & Kamide, 1999
3. Zwaan & Taylor, 2006
4. Taylor, Lev-Ari, & Zwaan, 2008
5. Sato, Schafer, & Bergen, 2012
6. Zwaan et al., 2002
7. Allopenna, Magnuson, & Tanenhaus, 1998
8. Huettig & Altmann, 2004
9. Dahan & Tanenhaus, 2005
10. Zwaan & Taylor, 2006
11. Townsend & Bever, 2001
12. Kaup, Yaxley, Madden, Zwaan, & Lüdtke, 2007
13. Giora, Balaban, Fein, & Alkabets, 2004
14. Kaup, Lüdtke, & Zwaan, 2006
15. Funny story about spending the last fifteen years of my life studying meaning. This one time I was flying out of SFO and I happened to have a jar of home-made quince preserves in my carry-on. A TSA agent stopped me, saying that the quince preserves couldn't come aboard because no gels, liquids, or aerosols were allowed past the checkpoint. I asked him politely which of those quince preserves were: gel, liquid or aerosol, because they seemed a lot like fruit. His response, and I kid you not, was

"Sir, I'm not going to argue semantics with you." And I wanted to say, "But THAT'S WHAT I DO!"

Notes to Chapter 7

1. Holt & Beilock, 2006
2. Beilock, Lyons, Mattarella-Micke, Nusbaum, & Small, 2008
3. A design like this, where you sample from different populations, is technically a "quasi-experiment" because you're not actually manipulating the relevant independent variable. Studies of individual differences, especially when those differences are purportedly due to long-term experience, always suffer from this difficulty.
4. Wassenburg & Zwaan, 2010
5. Galton, 1883
6. Richardson, 1977
7. Soloman & Barsalou, 2004
8. Carmichael, Hogan, & Walter, 1932
9. Shergill et al., 2002
10. Aziz-Zadeh, Cattaneo, Rochat, & Rizzolatti, 2005
11. Jorgensen, Lee, & Agabont, 2003
12. Blajenkova, Kozhevnikov, & Motes, 2006
13. Adapted from Kozhevnikov, Kosslyn, & Shephard, 2005
14. Look for an umbrella.
15. I just copied and pasted it. I honestly still haven't read the whole passage. I couldn't take it. Does that make me a verbalizer?
16. Dils & Borodistky, 2010
17. Winawer, Huk, & Boroditsky, 2010

Notes to Chapter 8

1. Mauss, 1936
2. Dennison & Bergen, 2010
3. Daniels & Bright, 1996
4. United Nations Development Program, 2009, Table H
5. Chatterjee et al., 1999
6. Maass, & Russo, 2003
7. Whorf, 1956
8. Majid, Bowerman, Kita, Haun, & Levinson, 2004
9. Boroditsky & Gaby, 2010
10. Kay, Berlin, Maffi, Merrifield, & Cook, 2010
11. Boroditsky, 2011

Notes to Chapter 9

1. Lakoff, 1993; Lakoff & Johnson, 1980
2. White's Residential and Family Services, 2009, retrieved October 5, 2011, from http://www.whiteskids.org/news/columbus-five-years.pdf
3. Lightbeam, 2005, retrieved October 5, 2011, from http://www.jref.com/forum/all-things-japanese-26/japans-closed-society-vs-economic-troubles-21068/

4. Ramsey, 2011; http://fayobserver.com/articles/2011/09/25/1123535?sac=Mil, accessed October 5, 2011

5. Kissell, 2010, retrieved October 5, 2011, from http://www.yourdailyjournal.com /view/full_story/7026031/article-Farmers-are-the-backbone-of-our-society?instance =search_results

6. Wallström, 2010, retrieved October 5, 2011, from http://bigthink.com/ideas/24674 &sa=U&ei=tErmTsuTFaX02QW5gem7BA&ved=0CBMQtwIwAQ&usg=AFQjC NHnL2_QyyXbFC5OvnzqvondwVSJww

7. Tony Blair Faith Foundation, 2011, retrieved October 5, 2011, from http://www .tonyblairfaithfoundation.org/blog/entry/a-healthy-society-requires-an-ongoing -dialogue-between-faith-and-reason/&sa=U&ei=W0vmTu-dJOrCsQKd792dBg &ved=0CCkQFjAHOAo&usg=AFQjCNFhyOU39NiQp6S4EH5JJyDslmJchg

8. Gibbs, Bogdanovich, Sykes, & Barr, 1997

9. Wilson & Gibbs, 2007

10. Stefanowitsch, 2004

11. Tseng, Hu, Han, & Bergen, 2005

12. Aziz-Zadeh & Damasio, 2008

13. Raposo, Moss, Stamatakis & Tyler, 2009

14. Boulenger, Hauk, & Pulvermüller, 2009

15. Bowdle, & Gentner, 2005

16. Desai, Binder, Conant, Mano, & Seidenberg, 2012

17. Bergen et al., 2007

18. Glenberg & Kaschak, 2002

19. Casasanto & Boroditsky, 2008

20. If you think about this claim for a moment, you might come up with the standard objection—that you can very well talk about the trip from San Diego to Los Angeles as a *two hour drive*, and isn't that an example of using time language to describe distance? This is actually an ambiguous case, in two ways. First, we don't know if you're using time language to talk about distance (by *two hour*, you mean 130 miles) or whether you're really literally talking about the temporal duration of the trip. I suspect you usually mean the latter. And even if you are really trying to describe a distance using language for time, it's not clear that that's metaphorical and not, as we'll see shortly, metonymic. Metaphor can be used not only when, as in this case, the two domains co-occur, but also when they do not. That contrasts with metonymy. Calling something a two-hour drive only makes sense in the context in which someone makes the real or imaginary drive. That's different from time-as-space metaphorical language, which makes sense despite the fact that there's no actual co-occurrence of the two—you can say that spring is around the corner even though there's never any physical corner involved. So *two hour drive*, if it's not literal (which I think it usually is) still probably isn't metaphorical.

21. Williams & Bargh, 2008

22. Zhong & Leonardelli, 2008

23. Zhong & Liljenquist, 2006

24. Matlock, 2004

25. Richardson & Matlock, 2007

Notes to Chapter 10

1. Lindsay, 2003

2. Kaschak et al., 2005

3. Glenberg, Sato, & Cattaneo, 2008; Meteyard, Zokaei, Bahrami, & Vigliocco, 2008
4. Damasio et al., 2001
5. Kemmerer, 2005
6. Shapiro et al., 2005
7. Damasio & Tranel, 1993
8. Shapiro, Pascual-Leone, Mottaghy, Gangitano, & Caramazza, 2001
9. Damasio, Grabowski, Tranel, Hichwa, & Damasio, 1996
10. Narayanan, 1997

Notes to Chapter 11

1. This semantic turn is on display in a lot of places. Perhaps most relevantly to you is how the Internet works. When search engines were first created in the early 1990s, they searched exclusively on the basis of the form of web pages. You typed some keywords into Lycos or Altavista, and they would search for web pages that had those exact words on them—that exact same form. But, in the past decade, search has changed a little bit. What if a search engine were able to scour not only the forms of web pages but also what those forms meant? What if, for example, a search engine knew what a web page was about and what you could do with it? That could potentially improve search a lot. Tim Berners-Lee, the MIT computer scientist credited with creating the World Wide Web, has called this idea the *semantic web*. The development of the semantic web has been slow, in part because, whereas form is easily to access, categorize, and process, meaning is much less so. To search web pages for their form (the characters that appear on them) is easy. But to search them for what those characters mean is no mean feat. That more challenging task is the direction that computing is going. And there's no reason to believe it won't happen in our lifetimes. The Internet has turned to meaning.
2. Pilley & Reid, 2011
3. Kellogg & Kellogg, 1933
4. It seems that many people are swayed by a compelling intuition that we are so vastly different from other animals that there couldn't in principle be any set of mutations that would produce us. This intuition is at work in creation myths in everything from traditional and religious thought to science fiction like Star Trek and Battlestar Galactica. But convenient as it might be if humans were planted on earth by a deity or by spacefaring protohumanoid explorers, all the evidence points to us coming about in the same, slow, incremental, mechanistic way as every other living thing we know about.
5. Brooks & Peever, 2008
6. Barsalou, 2005
7. Fias & Fischer, 2005
8. Dehaene, Bossini, & Giraux, 1993; Fias, Brysbaert, Geypens, & d'Ydewalle, 1996
9. Loetscher, Bockisch, Nicholls, & Brugger, 2010
10. Umiltà, Priftis, & Zorzi, 2009
11. Casasanto, 2008
12. Jostmann, Lakens, & Schubert, 2009

Notes to Epilogue

1. Alm & Nilsson, 1995; Lamble, Kauranen, Laakso, & Summala, 1999; Lee, McGehee, Brown, & Reyes, 2002; Levy, Pashler, & Boer, 2006; Strayer, Drews, & Johnston, 2003

2. Brookhuis, de Vries, & de Waard, 1991
3. Becic et al., 2010
4. Strayer & Drews, 2003; Strayer, Drews, & Crouch, 2006; Strayer et al., 2003
5. Atchley, Dressel, Jones, Burson & Marshall, 2011; Bergen, Medeiros-Ward, Wheeler, Drews, & Strayer, 2012

BIBLIOGRAPHY

Aginsky, V. & Tarr, M. J. (2000). How are different properties of a scene encoded in visual memory? *Visual Cognition, 7,* 147–162.

Allopenna, P. D., Magnuson, J. S., & Tanenhaus, M. K. (1998). Tracking the time course of spoken word recognition using eye movements: evidence for continuous mapping models. *Journal of Memory and Language, 38,* 419–439.

Alm, H. & Nilsson, L. (1995). The effects of a mobile telephone task on driver behaviour in a car following situation. *Accident Analysis & Prevention, 27,* 707–715.

Altmann, G. T. M. & Kamide, Y. (1999). Incremental interpretation at verbs: Restricting the domain of subsequent reference. *Cognition, 73,* 247–264.

Arnold, J., Wasow, T., Losongco, T., & Ginstrom, R. (2000). Heaviness vs. newness: The effects of structural complexity and discourse status on constituent ordering. *Language, 76,* 28–55.

Atchley, P., Dressel, J., Jones, T. C., Burson, R. A., & Marshall, D. (2011). Talking and driving: applications of crossmodal action reveal a special role for spatial language. *Psychological Research, 75*(6), 525–534.

Aziz-Zadeh, L., Cattaneo, L., Rochat, M., & Rizzolatti, G. (2005). Covert speech arrest induced by rTMS over both motor and nonmotor left hemisphere frontal sites. *Journal of Cognitive Neuroscience, 17,* 928–938.

Aziz-Zadeh, L. & Damasio, A. (2008). Embodied semantics for actions: Findings from functional brain imaging. *Journal of Physiology (Paris), 102,* 35–39.

Bailey, D. (1997). *When Push Comes to Shove: A Computational Model of the Role of Motor Control in the Acquisition of Action Verbs.* (Unpublished doctoral dissertation). University of California, Berkeley.

Barsalou, L. W. (1999). Perceptual symbol systems. *Behavioral and Brain Sciences, 22,* 577–609.

Barsalou, L. W. (2005). Continuity of the conceptual system across species. *Trends in Cognitive Sciences, 9,* 309–311.

Becic, E., Dell, G. S., Bock, K., Garnsey, S. M., Kubose, T., & Kramer, A. F. (2010). Driving impairs talking. *Psychonomic Bulletin & Review, 17,* 15–21.

Beilock, S. L., Lyons, I. M., Mattarella-Micke, A., Nusbaum, H. C., & Small, S. L. (2008). Sports experience changes the neural processing of action language. In *Proceedings of the National Academy of Sciences (USA), 105,* 13269–13273.

Bergen, B. & Chang, N. (2005). Embodied construction grammar in simulation-based language understanding. In J.-O. Östman & M. Fried (Eds.), *Construction grammars: Cognitive grounding and theoretical extensions* (pp. 147–190). Amsterdam: Benjamins.

Bergen, B., Lindsay, S., Matlock, T., & Narayanan, S. (2007). Spatial and linguistic aspects of visual imagery in sentence comprehension. *Cognitive Science, 31,* 733–764.

Bergen, B., Medeiros-Ward, N., Wheeler, K., Drews, F., & Strayer, D. (2012). The crosstalk hypothesis: Why language interferes with driving. *Journal of Experimental Psychology: General.* Manuscript in preparation.

Bergen, B. & Wheeler, K. (2005). Sentence understanding engages motor processes. In *Proceedings of the Twenty-Seventh Annual Conference of the Cognitive Science Society.*

Bergen, B. & Wheeler, K. (2010). Grammatical aspect and mental simulation. *Brain & Language, 112,* 150–158.

Blajenkova, O., Kozhevnikov, M., & Motes, M. A. (2006). Object-spatial imagery: a new self-report imagery questionnaire. *Applied Cognitive Psychology, 20,* 239–263.

Blanz, V., Tarr, M. J., & Bülthoff, H. H., (1999). What object attributes determine canonical views? *Perception, 28,* 575–600.

Borghi, A. M., Glenberg, A. M., & Kaschak, M. P. (2004). Putting words in perspective. *Memory and Cognition, 32,* 863–873.

Boroditsky, L. (2011). How language shapes thought. *Scientific American,* February 2011.

Boroditsky, L. & Gaby, A. (2010). Remembrances of times east: Absolute spatial representations of time in an Australian aboriginal community. *Psychological Science, 21*(11), 1635–1639.

Boroditsky, L., Schmidt, L. A., & Phillips, W. (2003). Sex, syntax and semantics. In D. Gentner & S. Goldin-Meadow (Eds.), *Language in mind: Advances in the study of language and thought* (pp. 61–79). Cambridge, MA: MIT Press.

Borreggine, K. L. & Kaschak, M. P. (2006). The action-sentence compatibility effect: it's all in the timing. *Cognitive Science, 30,* 1097–1112.

Boulenger, V., Hauk, O., & Pulvermüller, F. (2009). Grasping ideas with the motor system: Semantic somatotopy in idiom comprehension. *Cerebral Cortex, 19,* 1905–1914.

Bowdle, B. & Gentner, D. (2005). The career of metaphor. *Psychological Review, 112*(1), 193–216.

Brookhuis, K. A., de Vries, G., & de Waard, D. (1991). The effects of mobile telephoning on driving performance. *Accident Analysis & Prevention, 23,* 309–316.

Brooks, P. L. & Peever, J. (2008). Unraveling the mechanisms of REM sleep atonia. *Sleep, 31,* 1492–1497.

Brunye, T. T., Ditman, T., Mahoney, C. R., Augustyn, J. S., & Taylor, H. A. (2009). When you and I share perspectives: Pronouns modulate perspective-taking during narrative comprehension. *Psychological Science, 20,* 27–32.

Bub, D. N., Masson, M. E. J, & Cree, G. S. (2008). Evocation of functional and volumetric gestural knowledge by objects and words. *Cognition, 106*(1), 27–58.

Buccino, G., Riggio, L., Melli, G., Binkofski, F., Gallese, V., & Rizzolatti, G. (2005). Listening to action-related sentences modulates the activity of the motor system: a combined TMS and behavioral study. *Cognitive Brain Research, 24,* 355–363.

Burgund, E. D. & Marsolek, C. (2000). Viewpoint-invariant and viewpoint-dependent object recognition in dissociable neural subsystems. *Psychonomic Bulletin & Review, 7*(3), 480–489.

Carmichael, L., Hogan, H. P., & Walter, A. A. (1932). An experimental study of the effect of language on the reproduction of visually perceived forms. *Journal of Experimental Psychology, 15,* 73–86.

Carreiras, M., Carriedo, N., Alonso, M. A., & Fernández, A. (1997). The role of verb tense and verb aspect in the foregrounding of information during reading. *Memory & Cognition, 25,* 438–446.

Casasanto, D. (2008). Similarity and proximity: When does close in space mean close in mind? *Memory & Cognition, 36,* 1047–1056.

Casasanto, D. & Boroditsky, L. (2008). Time in the mind: Using space to think about time. *Cognition, 106,* 579–593.

Chan, T. T. & Bergen, B. (2005). Writing direction influences spatial cognition. In *Proceedings of the Twenty-Seventh Annual Conference of the Cognitive Science Society.*

Chang, N., Gildea, D., & Narayanan, S. (1998). A dynamic model of aspectual composition. In *Proceedings of the Twentieth Annual Conference of the Cognitive Science Society.*

Chatterjee, A. (2001). Language and space: some interactions. *Trends in Cognitive Sciences, 5*(2), 55–61.

Chatterjee, A., Southwood, M. H., & Basilico, D. (1999). Verbs, events and spatial representations. *Neuropsychologia, 37*(4), 395–402.

Chomsky, N. (1955). *The logical structure of linguistic theory.* Chicago: University of Chicago Press.

Comrie, B. (1976). *Aspect.* Cambridge: Cambridge University Press.

Connell, L. (2007). Representing object colour in language comprehension. *Cognition, 102,* 476–485.

Cooper, L.A. (1975). Mental rotation of random two-dimensional shapes. *Cognitive Psychology, 7,* 20–43.

Cooper, L. A. & Shepard, R. N. (1973). Chronometric studies of the rotation of mental images. In W. G. Chase (Ed.), *Visual information processing.* New York: Academic Press.

Craver-Lemley, C. & Arterberry, M. E. (2001). Imagery-induced interference on a visual detection task. *Spatial Vision, 14,* 101–119.

Cutzu, F. & Edelman, S. (1994). Canonical views in object representation and recognition. *Vision Research, 34,* 3037–3056

Dahan, D. & Tanenhaus, M. (2005). Looking at the rope when looking for the snake: conceptually mediated eye movements during spoken-word recognition. *Psychonomic Bulletin & Review, 12,* 453–459.

Damasio, A. R. & Tranel, D. (1993). Nouns and verbs are retrieved with differently distributed neural systems. In *Proceedings of the National Academy of Sciences (USA), 90,* 4957–4960.

Damasio, H., Grabowski, T. J., Tranel, D., Hichwa, R. D., & Damasio, A. R. (1996). A neural basis for lexical retrieval. *Nature, 380*(6574), 499–505.

Damasio, H., Grabowski, T. J., Tranel, D., Ponto, L. L. B., Hichwa, R. D., & Damasio, A. R. (2001). Neural correlates of naming actions and of naming spatial relations. *NeuroImage, 13,* 1053–1064.

Daniels, P. T. & Bright, W. (Eds.). (1996). *The world's writing systems.* New York: Oxford University Press.

Dehaene, S., Bossini, S., & Giraux, P. (1993). The mental representation of parity and numerical magnitude. *Journal of Experimental Psychology: General, 122,* 371–396.

Denis, M. & Kosslyn, S. M. (1999). Scanning visual mental images: A window on the mind. *Cahiers de Psychologie Cognitive (Current Psychology of Cognition), 18*(4), 409–65.

Dennison, H. & Bergen, B. (2010). Language-driven motor simulation is sensitive to social context. In *Proceedings of the Thirty-Second Annual Meeting of the Cognitive Science Society*.

Desai, R. H., Binder, J. R., Conant, L. L., Mano, Q. R., & Seidenberg, M. S. (2012). The neural career of sensorimotor metaphors. *Journal of Cognitive Neuroscience, 23*(9), 2376–2386.

Dils, A. T. & Boroditsky, L. (2010). A visual motion aftereffect from understanding motion language. In *Proceedings of the National Academy of Sciences (USA), 107*, 16396–16400.

Dowty, D. R. (1977). Toward a semantic analysis of verb aspect and the English "imperfective" progressive. *Linguistics and Philosophy, 1*, 45–79.

Driskell, J., Copper, C., & Moran, A. (1994). Does mental practice enhance performance? *Journal of Applied Psychology, 79*, 481–492.

Edelman, S. & Bülthoff, H. H. (1992). Orientation dependence in the recognition of familiar and novel views of 3D objects. *Vision Research, 32*, 2385–2400.

Ehrsson, H. H., Geyer, S., & Naito, E. (2003). Imagery of voluntary movement of fingers, toes, and tongue activates corresponding body-part-specific motor representations. *Journal of Neurophysiology, 90*, 3304–3316.

Farah, M. J. (1985). Psychophysical evidence for a shared representational medium for visual images and percepts. *Journal of Experimental Psychology: General, 114*, 93–105.

Farah, M. J. (1989). Semantic and perceptual priming: How similar are the underlying mechanisms? *Journal of Experimental Psychology: Human Perception and Performance, 15*, 188–194.

Feldman, J. A. (2006). *From molecule to metaphor: A neural theory of language*. Cambridge, MA: MIT Press.

Fias, W., Brysbaert, M., Geypens, F., & d'Ydewall, G. (1996). The importance of magnitude information in numerical processing: evidence from the SNARC effect. *Mathematical Cognition, 2*, 95–110.

Fias, W. & Fischer, M. H. (2005). Spatial representation of numbers. In J. Campbell (Ed.), *Handbook of mathematical cognition* (pp. 43–54). New York: Psychology Press.

Finke, R. A. (1989). *Principles of mental imagery*. Cambridge, MA: MIT Press.

Fodor, J. (1975). *The language of thought*, Cambridge, MA: Harvard University Press.

Gallese, V., Fadiga, L., Fogassi, L., & Rizzolatti, G. (1996). Action recognition in the premotor cortex. *Brain, 119*(2), 593–609.

Galton, F. (1883). *Inquiries into human faculty and its development*. London: Dent.

Gibbs, R. W., Bogdanovich, J. M., Sykes, J. R., & Barr, D. J. (1997). Metaphor in idiom comprehension. *Journal of Memory and Language, 37*, 141–154.

Giora, R., Balaban, N., Fein, O., & Alkabets, I. (2004). Negation as positivity in disguise. In H. L. Colston & A. Katz (Eds.), *Figurative language comprehension: Social and cultural influences* (pp. 233–258). Mahwah, NJ: Erlbaum.

Glenberg, A. M. & Kaschak, M. P. (2002). Grounding language in action. *Psychonomic Bulletin & Review, 9*(3), 558–565.

Glenberg, A. M. & Robertson, D. (2000). Symbol grounding and meaning: A comparison of high-dimensional and embodied theories of meaning. *Journal of Memory and Language, 43*, 379–401.

Glenberg, A. M., Sato, M., & Cattaneo, L. (2008). Use-induced motor plasticity affects the processing of abstract and concrete language. *Current Biology, 18*, R290–R291.

Glenberg, A. M., Sato, M., Cattaneo, L., Riggio, L., Palumbo, D., & Buccino, G. (2008) Processing abstract language modulates motor system activity. *Quarterly Journal of Experimental Psychology, 61*, 905–919.

Goldberg, A. E. (1995). *Constructions: A construction grammar approach to argument structure*. Chicago: Chicago University Press

Hagoort, P. (2005). On Broca, brain, and binding: a new framework. *Trends in Cognitive Sciences, 9*, 416–423.

Halpern, A. R. & Zatorre, R. J. (1999). When that tune runs through your head: a PET investigation of auditory imagery for familiar melodies. *Cerebral Cortex, 9*, 697–704.

Hartgens, F. & Kuipers, H. (2004). Effects of androgenic-anabolic steroids in athletes. *Sports Medicine, 34*, 513–554.

Hauk, O., Johnsrude, I., & Pulvermüller, F. (2004). Somatotopic representation of action words in human motor and premotor cortex. *Neuron, 41*, 301–307.

Holt, L. E. & Beilock, S. L. (2006). Expertise and its embodiment: Examining the impact of sensorimotor skill expertise on the representation of action-related text. *Psychonomic Bulletin & Review, 13*, 694–701.

Horton, W. S. & Rapp, D. N. (2003). Occlusion and the accessibility of information in narrative comprehension. *Psychonomic Bulletin & Review, 10*(1), 104–109.

Huettig, F. & Altmann, G. T. M. (2004). The online processing of ambiguous and unambiguous words in context: Evidence from head-mounted eye-tracking. In M. Carreiras & C. Clifton (Eds.), *The on-line study of sentence comprehension: Eyetracking, ERP and beyond* (pp. 187–207). New York: Psychology Press.

Iacoboni, M., Woods, R. P., Brass, M., Bekkering, H., Mazziotta, J. C., & Rizzolatti, G. (1999). Cortical mechanisms of human imitation. *Science, 286*, 2526–2528.

Johansson, R., Holsanova, J., & Holmqvist, K. (2006). Pictures and spoken descriptions elicit similar eye movements during mental imagery, both in light and in complete darkness. *Cognitive Science, 30*(6), 1053–1079.

Jorgensen, C., Lee, D., & Agabont, S. (2003). Sub auditory speech recognition based on EMG/EPG signals. In *Proceedings of the International Joint Conference on Neural Networks*.

Jostmann, N. B., Lakens, D., & Schubert, T. W. (2009). Weight as an embodiment of importance. *Psychological Science, 20*, 1169–1174.

Kaschak, M. P. & Borreggine, K. L. (2008). Is long-term structural priming affected by patterns of experience with individual verbs? *Journal of Memory and Language, 58*, 862–878.

Kaschak, M. P. & Glenberg, A. M. (2000). Constructing meaning: The role of affordances and grammatical constructions in sentence comprehension. *Journal of Memory & Language, 43*, 508–529.

Kaschak, M. P., Madden, C. J., Therriault, D. J., Yaxley, R. H., Aveyard, M. E., Blanchard, A. A., & Zwaan, R. A. (2005). Perception of motion affects language processing. *Cognition, 94*, B79–B89.

Kaup, B., Lüdtke, J., & Zwaan, R.A. (2006). Processing negated sentences with contradictory predicates: Is a door that is open mentally closed? *Journal of Pragmatics, 38*, 1033–1050.

Kaup, B., Yaxley, R. H., Madden, C. J., Zwaan, R. A., & Lüdtke, J. (2007). Experiential simulations of negated text information. *Quarterly Journal of Experimental Psychology, 60*, 976–990.

Kay, P., Berlin, B., Maffi, L., Merrifield, W. R., & Cook, R. (2010). *World color survey*. Stanford, CA: Center for the Study of Language and Information.

Kellogg, W. N. & Kellogg, L. A. (1933). *The ape and the child: A comparative study of the environmental influence upon early behavior*. New York: Whittlesey House.

Kemmerer, D. (2005). The spatial and temporal meanings of English prepositions can be independently impaired. *Neuropsychologia, 43,* 797–806.

Kissell, L. (2010, April 9). Farmers are the backbone of our society. Retrieved October 5, 2011, from http://www.yourdailyjournal.com/view/full_story/7026031/article-Farmers-are-the-backbone-of-our-society?instance=search_results.

Kosslyn, S. M., Ball, T. M., & Reiser, B.J . (1978). Visual images preserve metric spatial information: Evidence from studies of image scanning. *Journal of Experimental Psychology: Human Perception and Performance, 4,* 46–60.

Kozhevnikov, M., Kosslyn, S., & Shephard, J. (2005). Spatial versus object visualizers: A new characterization of visual cognitive style. *Memory & Cognition, 33,* 710–726.

Kraemer, D. J. M., Macrae, C. N., Green, A. E., & Kelley, W. M. (2005). Musical imagery: Sound of silence activates auditory cortex. *Nature, 434,* 158.

Lakoff, G. (1987). *Women, fire and dangerous things: What categories reveal about the mind*. Chicago: University of Chicago Press.

Lakoff, G. (1993). The contemporary theory of metaphor. In A. Ortony (Ed.), *Metaphor and thought* (2nd ed., pp. 202–251). Cambridge: Cambridge University Press.

Lakoff, G. (2004). Don't think of an elephant! Know your values and frame the debate. White River Junction, VT: Chelsea Green.

Lakoff, G. & Johnson, M. (1980). *Metaphors we live by*. Chicago: University of Chicago Press.

Lamble, D., Kauranen, T., Laakso, M., & Summala, H. (1999). Cognitive load and detection thresholds in car following situations: Safety implications for using mobile (cellular) telephones while driving. *Accident Analysis and Prevention, 31,* 617–623.

Langacker, R. W. (1987). *Foundations of cognitive grammar: Volume 1. Theoretical prerequisites*. Stanford, CA: Stanford University Press.

Lee, J. D., McGehee, D. V., Brown, T. L., & Reyes, M. L. (2002). Collision warning timing, driver distraction, and driver response to imminent rear-end collisions in a high-fidelity driving simulator. *Human Factors, 44,* 314–334.

Levy, J., Pashler, H., & Boer, E. (2006). Central interference in driving: Is there any stopping the psychological refractory period? *Psychological Science, 17,* 228–235.

Libby, L. K. & Eibach, R. P. (2002). Looking back in time: Self-concept change affects visual perspective in autobiographical memory. *Journal of Personality and Social Psychology, 82,* 167–179.

Libby, L. K. & Eibach, R. P. (2007). How the self affects and reflects the content and subjective experience of autobiographical memory. In C. Sedikides & S. J. Spencer (Eds.), *The self* (pp. 75–91). New York: Psychology Press.

Lightbeam. (2005, December 31). Japan's closed society vs economic troubles [Msg 1]. Retrieved October 5, 2011, from http://www.jref.com/forum/all-things-japanese-26/japans-closed-society-vs-economic-troubles-21068/.

Lindsay, S. (2003). Visual priming of language comprehension. (Unpublished master's thesis). University of Sussex, Brighton, UK.

Loetscher, T., Bockisch, C. J., Nicholls, M. E., & Brugger, P. (2010). Eye position predicts what number you have in mind. *Current Biology, 20*(6), R264–R265.

Lopez, B. (1986). *Arctic dreams: Imaginations and desire in a northern landscape*. New York: Scribners.

Lotze, M., Montoya, P., Erb, M., Hülsmann, E., Flor, H., Klose, U., Birbaumer, N., & Grodd, W. (1999). Activation of cortical and cerebellar motor areas during executed and imagined hand movements: An fMRI study. *Journal of Cognitive Neuroscience,* *11*(5), 491–501.

Maass, A. & Russo, A. (2003). Directional bias in the mental representation of spatial events: Nature or culture? *Psychological Science, 14,* 296–301.

Madden, C. J. & Zwaan, R.A. (2003). How does verb aspect constrain event representations? *Memory & Cognition, 31,* 663–672.

Magliano, J. P. & Schleich, M. C. (2000). Verb aspect and situation models. *Discourse Processes, 29,* 83–112.

Maguire, E. A., Valentine, E. R., Wilding, J. M., & Kapur, N. (2003). Routes to remembering: the brains behind superior memory. *Nature Neuroscience, 6,* 90–95.

Majid, A., Bowerman, M., Kita, S., Haun, D., & Levinson, S. (2004). Can language restructure cognition? The case for space. *Trends in Cognitive Sciences, 8,* 108–114.

Masson, M. E. J., Bub, D. N., & Warren, C. M. (2008). Kicking calculators: Contribution of embodied representations to sentence comprehension. *Journal of Memory and Language, 59,* 256–265.

Matlock, T. (2004). Fictive motion as cognitive simulation. *Memory and Cognition, 32,* 1389–1400.

Mauss, M. (1936). "Les techniques du corps." *Journal de Psychologie, 32*(3–4), 365–386.

Mellet, E., Bricogne, S., Crivello, F., Mazoyer, B., Denis, M., & Tzourio-Mazoyer, N. (2002). Neural basis of mental scanning of a topographic representation built from a text. *Cerebral Cortex, 12,* 1322–1330.

Meteyard, L., Zokaei, N., Bahrami, B., & Vigliocco, G. (2008). Visual motion interferes with lexical decision on motion words. *Current Biology, 18,* R732–R733.

Narayanan, S. (1997). KARMA: Knowledge-based active representations for metaphor and aspect. (Unpublished doctoral dissertation). University of California, Berkeley.

Nigro, G. & Neisser, U. (1983). Point of view in personal memories. *Cognitive Psychology, 15,* 467–482.

Nyberg, L., Petersson, K.-M., Nilsson, L.-G., Sandblom, J., Åberg, C., & Ingvar, M. (2001). Reactivation of motor brain areas during explicit memory for actions. *NeuroImage, 14,* 521–528.

Olin, C. H. (1910). *Phrenology.* Philadelphia: Kessinger.

Palmer, S., Rosch, E., & Chase, P. (1981). Canonical perspective and the perception of objects. In J. Long & A. Baddeley (Eds.), *Attention and Performance IX* (pp. 135–151). Hillsdale, NJ: Erlbaum.

Pecher, D., Zeelenberg, R., & Barsalou, L.W. (2003). Verifying different modality properties for concepts produces switching costs. *Psychological Science, 14,* 119–124.

Penfield, W. (1958). Some mechanisms of consciousness discovered during electrical stimulation of the brain. In *Proceedings of the National Academy of Sciences (USA), 44,* 51–66.

Perky, C. W. (1910). An experimental study of imagination. *American Journal of Psychology, 21,* 422–452.

Pfanz, H, Aschan, G., Langenfeld-Heyser, R., Wittmann, C., & Loose, M. (2002). Ecology and ecophysiology of tree stems: Corticular and wood photosynthesis. *Naturwissenschaften 89*(4), 147–162.

Pinker, S. (1994). *The language instinct: How the mind creates language.* New York: HarperCollins.

Pilley, J. W. & Reid, A. K. (2011). Border collie comprehends object names as verbal referents. *Behavioral Processes, 86,* 184–195.

Porro, C. A., Francescato, M. P., Cettolo, V., Diamond, M. E., Baraldi, P., Zuiani, C., Bazzocchi, M., & di Prampero, P. E. (1996). Primary motor and sensory cortex activation during motor performance and motor imagery: A functional magnetic resonance imaging study. *Journal of Neuroscience, 16*, 7688–7698.

Pulvermüller, F., Härle, M., & Hummel, F. (2001). Walking or talking? Behavioral and neurophysiological correlates of action verb processing. *Brain and Language, 78*, 143–168.

Ramachandran, V. (2006, January 10). Mirror neurons and the brain in the vat. Retrieved from http://www.edge.org/3rd_culture/ramachandran06/ramachandran06_index.html

Ramsey, J. (2011, September 25). Collateral damage: War veterans struggle to fit back into society. Retrieved October 5, 2011, from http://fayobserver.com/articles/2011/09/25/1123535?sac=Mil

Raposo, A., Moss, H. E., Stamatakis, E. A., & Tyler, L. K. (2009). Modulation of motor and premotor cortices by actions, action words and action sentences. *Neuropsychologia, 47*, 388–396.

Richardson, A. (1977). Verbalizer-visualizer: A cognitive style dimension. *Journal of Mental Imagery, 1*, 109–125.

Richardson, D. C. & Matlock, T. (2007). The integration of figurative language and static depictions: an eye movement study of fictive motion. *Cognition, 102*, 129–138.

Richardson, D. C., Spivey, M. J., Barsalou, L. W., & McRae, K. (2003). Spatial representations activated during real-time comprehension of verbs. *Cognitive Science, 27*, 767–780.

Rizzolatti, G., Fadiga, L., Gallese, V., & Fogassi, L. (1996). Premotor cortex and the recognition of motor actions. *Cognitive Brain Research, 3*, 131–141.

Roediger, H. L., III. (1980). Memory metaphors in cognitive psychology. *Memory and Cognition, 8*, 231–246

Ruby, P. & Decety, J. (2001). Effect of subjective perspective taking during simulation of action: A PET investigation of agency. *Nature Neuroscience, 4*, 546–550.

Russell, B. (1903). *Principles of Mathematics*. Cambridge: Cambridge University Press.

Sato, M., Schafer, A., & Bergen, B. (2012). Mental representations of object shape change incrementally during sentence processing. Language and Cognition.

Searle, J. (1999) The Chinese room. In R. A. Wilson & F. Keil (Eds.), *The MIT encyclopedia of the cognitive sciences* (pp. 115–116). Cambridge, MA: MIT Press.

Segal, S. J. (1972). Assimilation of a stimulus in the construction of an image: The Perky effect revisited. In P. W. Sheehan (Ed.), *The function and nature of imagery* (pp. 203–230). New York: Academic Press.

Shapiro, K., Mottaghy, F. M., Schiller, N. O., Poeppel, T. D., Fluss, M. O., Muller, H. W., Caramazza, A., & Krause, B.J. (2005). Dissociating neural correlate for verbs and nouns. *Neuroimage, 24*, 1058–1067.

Shapiro, K.A., Pascual-Leone, A., Mottaghy, F.M., Gangitano, M., & Caramazza, A. (2001). Grammatical distinctions in the left frontal cortex. *Journal of Cognitive Neuroscience, 13*(6), 713–720.

Shepard, R. N. (1975). Form, formation and transformation of internal representations. In R. Solso (Ed.), *Information processing and cognition: The Loyola symposium*. Potomac, MD: Erlbaum.

Shepard, R. N. & Metzler, J. (1971). Mental rotation of three-dimensional objects. *Science, 171*, 701–703.

Shergill, S., Brammer, M., Fukuda, R., Bullmore, E., Amaro, E., Murray, R., & McGuire, P. (2002). Modulation of activity in temporal cortex during generation of inner speech. *Human Brain Mapping, 16*, 219–27.

Siakaluk, P. D., Pexman, P. M., Aguilera, L., Owen, W. J., & Sears, C. R. (2008). Evidence for the activation of sensorimotor information during visual word recognition: The body-object interaction effect. *Cognition, 106*, 433–443.

Solomon, K. O. & Barsalou, L. W. (2004). Perceptual simulation in property verification. *Memory and Cognition, 32*, 244–259.

Spectator, December 21, 1901. p. 984

Spivey, M. J. & Geng, J. J. (2001). Oculomotor mechanisms activated by imagery and memory: Eye movements to absent objects. *Psychological Research, 65*, 235–241.

Stanfield, R. A. & Zwaan, R. A. (2001). The effect of implied orientation derived from verbal context on picture recognition. *Psychological Science, 12*, 153–156.

Stefanowitsch, A. (2004). HAPPINESS in English and German: A metaphorical-pattern analysis. In M. Achard & S. Kemmer (Eds.), *Language, culture, and mind* (pp. 137–149). Stanford, CA: Center for the Study of Language and Information.

Stirling, I. (1974). Midsummer observations on the behavior of wild polar bears. (Ursus maritimus). *Canadian Journal of Zoology, 52*, 1191–1198.

Strayer, D. L. & Drews, F. A. (2003). Effects of cell phone conversations on younger and older drivers. In *Proceedings of the Human Factors and Ergonomics Society Forty-Seventh Annual Meeting.*

Strayer, D. L., Drews, F. A., & Crouch, D. J. (2006). Comparing the cellphone driver and the drunk driver. *Human Factors, 48*, 381–391.

Strayer, D. L., Drews, F. A., & Johnston, W. A. (2003). Cell phone-induced failures of visual attention during simulated driving. *Journal of Experimental Psychology: Applied, 9*, 23–32.

Strohecker, C. (2000). Cognitive zoom: From object to path and back again. In Freksa, C., Brauer, W., Habel, C., & Wender, K. F. (Eds.), *Spatial cognition II—Integrating abstract theories, empirical studies, formal methods, and practical applications* (pp. 1–15). Berlin: Springer.

Svensson, P. (1998). *Number and countability in English nouns: An embodied model.* Uppsala, SE: Swedish Science Press.

Talmy, L. (2000). Toward a cognitive semantics. In *Concept structuring systems: Volume 1.* Cambridge, MA: MIT Press.

Tarr, M. J. (1995). Rotating objects to recognize them: A case study on the role of viewpoint-dependency in the recognition of three dimensional objects. *Psychonomic Bulletin & Review, 2*, 55–82.

Tarr, M. & Pinker, S. (1989). Mental rotation and orientation-dependence in shape recognition. *Cognitive Psychology, 21*, 233–282.

Taylor, L. J., Lev-Ari, S., & Zwaan, R. A. (2008). Inferences about action engage action systems. *Brain and Language, 107*(1), 62–67.

Tettamanti, M., Buccino, G., Saccuman, M. C., Gallese, V., Danna, M., Scifo, P., Fazio, F., Rizzolatti, G., Cappa, S. F., & Perani, D. (2005). Listening to action-related sentences activates fronto-parietal motor circuits. *Journal of Cognitive Neuroscience, 17*, 273–281.

Tony Blair Faith Foundation. (2011). A healthy society requires an ongoing dialogue between faith and reason. Retrieved October 5, 2011, from http://www.tonyblairfaith foundation.org/blog/

Townsend, D. & Bever, T. G. (2001). *Sentence comprehension: The integration of habits and rules*. Cambridge, MA: MIT Press.

Treisman, A. (1996). The binding problem. *Current Opinions in Neurobiology, 6*, 171–178.

Trueswell, J. C., Tanenhaus, M. K., & Garnsey S. M. (1994). Semantic influences on parsing: Use of thematic role information in syntactic ambiguity resolution. *Journal of Memory and Language, 33*, 285–318.

Tseng, M. & Bergen, B. (2005). Lexical processing drives motor simulation. In *Proceedings of the Twenty-Seventh Annual Conference of the Cognitive Science Society*.

Tseng, M., Hu, Y., Han, W.-W., & Bergen, B. (2005). "Searching for happiness" or "full of joy"? Source domain activation matters. In *Proceedings of the Thirty-First Annual Meeting of the Berkeley Linguistics Society*.

Umiltà, C., Priftis, K., & Zorzi, M. (2009). The spatial representation of numbers: Evidence from neglect and pseudoneglect. *Experimental Brain Research, 192*, 561–569.

United Nations Development Program (2009). *Human development report: Overcoming barriers: Human mobility and development (Table H)*. New York: Palgrave Macmillan.

Vandenbeld, L. A. & Rensink, R. A. (2003). The decay characteristics of size, color, and shape information in visual short-term memory. *Journal of Vision, 3*, 682.

Vogeley, K. & Fink, G. R. (2003). Neural correlates of first-person-perspective. *Trends in Cognitive Sciences, 7*, 38–42.

Wallström, M. (2010, October 17). How sexual violence disempowers women and cripples society. Retrieved October 5, 2011, from http://bigthink.com/ideas/24674&sa =U&ei=tErmTsuTFaX02QW5gem7BA&ved=0CBMQtwIwAQ&usg=AFQjCNHnL2 _QyyXbFC5OvnzqvondwVSJww

Wassenburg, S. I. & Zwaan, R. A. (2010). Readers routinely represent implied object rotation: The role of visual experience. *Quarterly Journal of Experimental Psychology, 63*, 1665–1670.

Weinberg, R. (2008). Does imagery work? Effects on performance and mental skills. *Journal of Imagery Research in Sport and Physical Activity, 3*(1), 1–21.

Wheeler, M. E., Petersen, S. E., & Buckner, R. L. (2000). Memory's echo: Vivid remembering reactivates sensory-specific cortex. In *Proceedings of the National Academy of Sciences, (USA)*, 97(20), 11125–11129.

White's Residential and Family Services. (2009, October 28). News release and photo opportunity: Columbus office celebrates five years of regional impact. Retrieved October 5, 2011, from http://www.whiteskids.org/news/columbus-five-years.pdf

Whorf, B. L. (1956). *Language, thought, and reality: Selected writings of Benjamin Lee Whorf*. John B. Carroll (Ed.). Cambridge, MA: MIT Press.

Wilcox, T. (1999). Object individuation: Infants' use of shape, size, pattern, and color. *Cognition, 72*, 125–166.

Williams, L. E. & Bargh, J. A. (2008). Experiencing physical warmth influences interpersonal warmth. *Science, 322*, 606–607.

Wilson, N. L. & Gibbs, R.W., Jr. (2007). Real and imagined body movement primes metaphor comprehension. *Cognitive Science, 31*, 721–731.

Winawer, J., Huk, A., & Boroditsky, L. (2010) A motion aftereffect from visual imagery of motion. *Cognition, 114*(2), 276–284.

Winter, B. & Bergen, B. (2012). Language comprehenders represent object distance both visually and auditorily. *Language and Cognition, 4*(1), 1–16.

Woolfolk, R. L., Parrish, M. W., & Murphy, S. M. (1985). The effects of positive and negative imagery on motor skill performance. *Cognitive Therapy and Research, 9,* 335–341.

Yaxley, R. H. & Zwaan, R. A. (2007). Simulating visibility during language comprehension. *Cognition, 150,* 229–236.

Zhong, C. B. & Leonardelli, G. J. (2008). Cold and lonely: Does social exclusion feel literally cold? *Psychological Science, 19,* 838–842.

Zhong, C. B. & Liljenquist, K. (2006). Washing away your sins: Threatened morality and physical cleansing. *Science, 313,* 1451–1452.

Zwaan, R. A. (2004). The immersed experiencer: Toward an embodied theory of language comprehension. In B. H. Ross (Ed.), *The psychology of learning and motivation: Volume 43* (pp. 35–62). New York: Academic Press.

Zwaan, R. A. & Madden, C. J. (2005). Embodied sentence comprehension. In D. Pecher & R. A. Zwaan (Eds.), *Grounding cognition: The role of perception and action in memory, language, and thinking* (pp. 224–245). Cambridge: Cambridge University Press.

Zwaan, R. A., Madden, C. J., Yaxley, R. H., & Aveyard, M.E. (2004). Moving words: Dynamic mental representations in language comprehension. *Cognitive Science, 28,* 611– 619.

Zwaan, R. A., Stanfield, R. A., & Yaxley, R. H. (2002). Language comprehenders mentally represent the shapes of objects. *Psychological Science, 13,* 168–171.

Zwaan, R. A. & Taylor, L. J. (2006). Seeing, acting, understanding: motor resonance in language comprehension. *Journal of Experimental Psychology: General, 135,* 1–11.

PERMISSIONS

Figure 1 Superswine image by Frances Ajo, used with permission; Pigasus image used with permission of the Martha Heasley Cox Center for Steinbeck Studies.

Figure 2 Adapted from Shepard & Metzler (1971).

Figure 3 Adapted from Kosslyn, Ball, & Reiser (1978).

Figure 4 Reproduced by permission of the Literary Executor of the Estate of Wilder Penfield.

Figure 5 Adapted from an image by Selket.

Figure 6 Adapted by Frances Ajo, used with permission.

Figure 7 From Wheeler et al. (2000), used with permission from the National Academy of Sciences.

Figure 8 From Nyberg et al. (2001), reprinted with permission.

Figure 9 Adapted from an image by Selket.

Figure 11 Eagle in nest used with permission of Matthias Trischler. Eagle in the sky: Photo credit: NASA/Gary Rothstein.

Figure 13 From Yaxley & Zwaan (2007) used with permission of Elsevier.

Figure 14 Photo by Horia Varlan.

Figure 15 From Gallese, Fadiga, Fogassi, & Rizzolatti (1996), by permission of Oxford University Press.

Figure 16 Image by Frances Ajo, used with permission.

Figure 17 Adapted from Glenberg & Kaschak (2002).

Figure 18 From Bub, Masson, & Cree (2008), used with permission of Elsevier.

Figure 19 Image by Shweta Narayan, used with permission.

Figure 20 Adapted from Buccino et al. (2005).

Figure 21 From Pulvermüller, Härle, & Hummel (2001), used with permission of Elsevier.

Figure 22 Adapted from Pulvermüller et al. (2001).

Figure 23 Adapted from Tettamanti et al. (2005).

Figure 34 Adapted from Zwaan & Taylor (2006).

Figure 35 Intact egg licensed under the Creative Commons Attribution-Share Alike 3.0 Unported license by Wikimedia Commons user Sun Ladder. Cracked egg public domain, taken by David Benbennick.

Figure 37 Beetle and speaker images public domain; carriage adapted by author, beaker by author.

Figure 39 From Kaup, Luedke, & Zwaan (2006), with permission of Elsevier.

Figure 40 Adapted from Kaup et al. (2006) and Kaup et al. (2007).

Figure 41   From Beilock et al. (2008), with permission of the National Academy of
            Sciences.
Figure 42   Adapted from Wassenberg & Zwaan (2010).
Figure 44   Adapted from Carmichael et al. (1932).
Figure 45   Image adapted by Frances Ajo, used with permission.
Figure 47   Image public domain.
Figure 48   Photo by Alexander Cain, used with permission.
Figure 50   Image by Heeyeon Dennison, used with permission.
Figure 53   Adapted from Maass & Russo (2003).
Figure 54   Adapted from Stefanowitsch (2003).
Figure 57   From Casasanto & Boroditsky (2008), used with permission of Elsevier.
Figure 58   From Richardson & Matlock (2007), used with permission of Elsevier.
Figure 60   From Kaschak et al. (2005), used with permission of Elsevier.
Figure 61   From Shapiro et al. (2005), used with permission of Elsevier.

All other figures by author.

INDEX

Abstract concepts, 5–9, 207–209
    abstract thought, 50
    and metaphorical simulation, 210–211,
        215–216, 221
    in the absence of language, 211–214
    language about abstract concepts, 209–211,
        216, 240
    reasoning about, 240, 256
    See also Embodied simulation: and
        mathematics; Metaphor; Metaphorical
        simulation
Action planning, 157, 244–245
    and language, 74
    handshape, 83–84
    See also Motor imagery; Motor simulation;
        Motor system
Action-sentence compatibility, 79–83, 85, 113
    and abstract language, 210
    and culture, 179–181
    and event structure, 117
    and manual rotation, 86, 127–131
    and metaphorical language, 199–200
    See also Motor simulation
Argument structure constructions, 96–97, 104
    and embodied simulation, 106–108
    and meaning, 100–103
    caused-motion construction, 104–108
    ditransitive construction, 100–103,
        105–108
    principle of no synonomy, 104
    transfer of possession construction,
        100–103
    transitive construction, 96, 99, 102
    See also Grammar
Arm wrestling, 73–75
Audition
    auditory imagery, 34–35
    auditory simulation, 47, 166, 182–183, 224

auditory system, 16, 34–35, 164
    See also Music

Barsalou, Lawrence, 13
Binding problem, 93–94
Bonds, Barry, 23
Boroditsky, Lera, 170–172, 189, 212–214
Brain activity
    and experience, 156–159
    dissociation studies, 234–236
    during perception (*see* Perception: and
        brain activity)
    in memory, 42–44
    in metaphor, 203–206
    in music, 35–36
    in performance and imagery, 36–39
    in performance and recall, 44
    selective activation, 38–39, 44, 76, 91–92,
        235
    with different types of words, 44–45,
        235–237
    See also Cerebellum; Functional magnetic
        resonance imaging; Motor strip; Positron
        emission tomography; Reuse of brain
        systems; Transcranial magnetic
        stimulation

Cerebellum, 39
Chinese Room argument, 10
Cognitive science, 5, 13, 21, 250
    and language (*see* Language)
    See also Language comprehension;
        Meaning
Cognitive style, 162, 166, 169
    and motion after-effect (*see* Motion after-
        effect)
    object visualizers, 166–167, 172–173
    spatial visualizers, 166, 170, 172–173

291

86782

959 Lim, John, 1932-
Lim   Merchants of the mysterious East / John
        Lim. Montreal : Tundra, c1981.
          32 p. : ill.

     1. Asia, Southeastern - Social life and
customs. I. Title.
0887761305   1407082

6/he.

Merchants of the Mysterious East

First printing

Published in Canada by
Tundra Books of Montreal
Montreal, Quebec   H3G 1R4
ISBN 0-88776-130-5
Legal deposit, Third Quarter
Quebec National Library

Published in the United States by
Tundra Books of Northern New York
Plattsburgh, N.Y. 12901

Design by Michael M. Cutler
Transparencies by T.E. Moore
Typeset by Compoplus, Montreal
Printed by Herzig Somerville, Toronto

Printed in Canada

Published in Australia, New Zealand and Southeast Asia
by Childerset Pty. Ltd., Melbourne, Australia

Merchants of the Mysterious East   John Lim

Tundra Books

When Johnnie was a boy, the streets of Singapore were full of mystery, magic and color. The city was on an island and people came to live and work there from many parts of the world. Johnnie's grandparents came from China; others came from Malaya and Siam, from India and even from far-off England. Their languages added strange sounds to match the wonderful sights and smells.

On weekends, Johnnie was taken by his father to visit his grandmother who had a farm just outside the city. But other days, after school, he loved to explore the streets with his brother and sister and friends and to shop with his family. There was so much to look at and wonder about.

Johnnie is a man now and an artist. The Singapore he knew has become a modern city of apartment buildings and big factories. Most of the peddlers, fortune-tellers and little shopkeepers have disappeared. This is how he remembers them — with love and thanks for having made his boyhood such a beautiful and exciting time.

Tellers of fortunes, sellers of stories

The people of Singapore loved stories of the past and predictions for the future. So those who made up stories or told fortunes were very popular.

The Astrologer

To Johnnie, the most mysterious place in all Singapore was the Astrologer's office. He never got to visit it, but he liked to peek inside. He knew his father had gone there shortly after Johnnie was born to ask the Astrologer what kind of life lay ahead for his son.

On the wall behind the Astrologer was a chart covering the twelve-year cycle of zodiac signs. Each year had its own animal sign. Johnnie was born under the sign of the monkey. On other walls were charts of the heavens with planets and stars.

The Astrologer studied the position of the planets on the day and at the exact minute Johnnie was born, then told his father what name to give him and predicted the boy would travel far and be happy. The prediction was supposed to be a secret, kept only to the adults. But Johnnie's father announced the good news joyfully to all the family — and later to Johnnie when he was old enough to understand.

If the Astrologer predicted any bad things for him, his father did not tell him. So Johnnie grew up expecting only good things.

he Astrologer   *John Lim*

The Storyteller

John Li

The Storyteller

The Storyteller did not come around the same day every week, but everyone seemed to know when he would be in the neighborhood. On such evenings after supper, Johnnie and his family walked to the corner where a stool and table had been set up.

Sometimes the Storyteller read from books, changing his voice to suit the character speaking and stopping often to add details from his own imagination. But Johnnie liked best the stories told from memory, for then the Storyteller acted everything out. And what an actor he was! He ran in circles, pretending he was being chased. He swooped down from his table, like a bird landing. He fought off imaginary villains. When he pointed to the sky with his walking stick, everyone looked up, scared and excited, expecting to see the creature he pretended was there.

When the stories came to an end, the grownups put coins on the table to thank the Storyteller. All the way home, Johnnie acted out his favorite parts, imitating the voice and movements of the Storyteller.

The Embroidery Shop

To Johnnie the Embroidery Shop was also a place for stories. While his mother chose patterns to be embroidered on dresses, blouses and handkerchiefs and discussed the color of silk thread to be used, Johnnie wandered around on his own. The walls were a picture gallery of tapestries showing chow dogs, phoenixes, dragons and strange flowers. He had only to stand and watch the women at work on a tapestry for them to stop sewing and tell him stories of the figures they were embroidering.

The Fortuneteller

The sound of bamboo sticks rattling against a bamboo vase brought Johnnie running from his house, for it meant the Fortuneteller was passing by.

Johnnie was more interested in the bird he carried than in the fortunes. In return for a few pennies, the Fortuneteller chanted a secret prayer for good luck, then shook the bamboo vase in front of the birdcage. The bird stuck its beak out and picked a stick from the vase. The fortune was written on the stick, and the old man read it aloud very seriously. He then took a seed from the tiny pouch on his belt and paid the bird for its help. Since all the fortunes were happy, he was a welcome visitor to the street.

The Embroidery Shop

John Li

e Fortuneteller

John Lim

Peddlers of novelties, notions and necessities

You didn't have to go looking for many of the trades-
men who tried to make you pretty, handsome or
happy. They came right to the house.

The Parasol Vendor

Singapore is in the tropics and people must protect
themselves from the very hot sun. Working people
wore wide-brimmed hats woven of bamboo and cane
and women like Johnnie's mother who liked to look
pretty used parasols. The Parasol Vendor who drove
his bicycle cart through the streets sold both.

Parasols are small sun-umbrellas usually made of
lacquered paper with bamboo frames; expensive
models are made from silk with metal frames. All are
brightly colored with flowers, birds or butterflies
painted on them by hand.

Every time Johnnie's mother got a new dress, she
bought a parasol to match it. The Parasol Vendor was
a patient man who opened one parasol after another
until she found the right color and pattern. Johnnie's
sister also liked the Parasol Vendor because he carried
smaller parasols for little girls, so they could go
walking with their mothers in the sun.

e Parasol Vendor

John Lim

The Knickknack Vendor

John Li

The Knickknack Vendor

The Knickknack Vendor pretended to perform a useful service and he did sell such necessities as chopsticks, flour sifters, oriental frying pans called *woks* and needles and thread. But that was not the reason mothers and children rushed out of the house and gathered round him several times a week when he passed.

What he really sold was surprise, for he always had new things to show and sell. Dishes and china figurines, fans and paper flowers for the women, and toys of all kinds for the children. There were dolls and windmills, flutes and tin whistles, paper balloons, feather cocks and glass marbles, also paper birds and butterflies that hung by strings from bamboo sticks and whistled when twirled. Since nothing cost more than a few pennies, Johnnie was always sure to get something.

Of all the knickknacks, one impressed Johnnie most, for it seemed pure magic. It was a small tightly wrapped package, and you had only to put it in water for it to open into a great cluster of paper flowers.

The Tailor

Of all the merchants who came to the house, the Tailor was the least popular with Johnnie because of all the waiting. He had to wait while his father and older brother were measured for their new suits before he was measured for his. Then he had to wait while his parents looked over the sample fabrics and made up their minds what he should have. Only then did they ask him if he liked it.

The final fitting was much better. It was in the Tailor's shop and the whole family strolled there on the appointed evening. There too he had to wait around, but at least he could watch the workers cutting and sewing. At last it all ended well, when they were given new suits nicely packaged to carry home.

The Beautician

The day the Beautician came to the house with her powders, potions and instruments, Johnnie usually decided to stay outdoors. Not that she wasn't interesting to watch, but once was enough.

Along with the latest styles, the Beautician spread the latest news. She talked nonstop as she heated long iron curlers on an open fire, curled the hair, then lacquered it into place. She also plucked eyebrows and removed moles. Her most unusual trick was to remove facial hair by coating the face with thick white powder and then rolling strands of string over it.

Johnnie liked to come home after the Beautician had been there and find his mother looking elegant and pretty. She was particularly pleased when he told her how nice she looked.

The Tailor

John L

The Beautician

John Lim

Frogs and fragrance, herbs and spices

Strange foods, strange smells and strange medicines.
The merchants of Singapore sold them all.

The Frog Vendor

When the monsoon season came and rains flooded
the fields around Singapore, the cry of the Frog
Vendor sounded through the streets.

At night he went into the fields to catch the frogs that
were always very plentiful after a rainstorm. The next
day he went from house to house selling them.
Johnnie's mother chose the frogs she bought very
carefully. The Vendor grabbed a handful through a
small opening in the top of his basket and dropped
them into a pail of water for her to examine and
choose from. Johnnie and his brother stood by waiting
and watching. A frog was sure to escape. Then they
both sprang after it to see who could catch it first.

The whole family liked frogs for supper. Each year
they looked forward to the call of the Frog Vendor and
the special treat he brought.

e Frog Vendor

John Lim

The Incense Shop

John Li

The Incense Shop

The most beautiful-smelling shop in all Singapore was the one that sold incense. The counters were crammed with boxes of joss sticks, and each had its own fragrance. A joss is a religious idol or statue and many people of Chinese origin have one in their home. Joss sticks are burned in front of it while offerings are made and prayers said — much as many Westerners use candles. Some joss sticks are six feet tall. These will burn for days to mark special occasions and fill the house with perfume. They are also burned as offerings in the temple.

The Incense Shop also sold large candles covered with flowers and dragons. Red is considered the color of good luck, and Johnnie noticed that all the packages of incense and all the candles in the shop were red.

The Herb Shop

The Herb Shop looked mysterious and smelled good, but Johnnie knew from experience that everything in it tasted just awful.

You went there because you weren't feeling well, but were not sick enough to need a doctor. The Herbalist listened, examined you, recommended a medicine and then set about preparing it. He was interesting to watch, if you could forget that you had to drink the stuff later. He put a piece of white paper on the counter and then chose herbs to put on it from the bottles, jars, packages and drawers around him. As he weighed each ingredient, sometimes crushing it with a mortar and pestle, he explained why it would make you feel better. Some of the more exotic herbs were for adults only, and these were never explained to Johnnie: rhinoceros horn, dried sea horse, or ginseng — a root in the shape of a man.

At home, the herbs ceased to smell nice as they were boiled, sometimes for hours, and the final taste was *always* worse than the smell.

The Spice Grinder

Whenever Johnnie went to market with his mother, he could tell he was close to the Spice Grinder even before he could see her, just from the wonderful smells.

She sat with her roots, plants, cuttings, berries and nuts spread out around her like flowers. When Johnnie's mother described what meat or fish she had bought for supper, the Spice Grinder suggested a blend of spices to go with it. Her hands moved quickly, gathering the spices, crushing and grinding them with a roller and stone. Chillies, ginger, lemongrass, pepper cloves, nutmeg, anise seed — as each spice was added, the aroma changed and got stronger.

Finally, the mixture was finished. The Spice Grinder wrapped it, damp and pungent, in a banana leaf and handed it to Johnnie's mother with a wish for an enjoyable meal.

The Herb Shop

John Li

he Spice Grinder

John Lim

Flight and fire, sound and song

The tellers of stories and fortunes seemed to pull their ideas from the air, but to Johnnie the people of Singapore seemed to put far more things *into* the air than they took out of it. No wonder everyone was always looking up.

The Bird Shop

Birds came from the sky, but many also went back to it in a very beautiful way. In the East birds were the most popular housepets, and the Bird Shop was always busy and noisy, and somehow magical to Johnnie. Bright yellow canaries sang, pure white doves cooed, parrots chattered and love birds cuddled. The cages were handmade of fine bamboo, often very intricate. Many families took such pride in their birds they had special cages made to order.

They also took their birds out "walking" in the evening, just as Westerners walk their dogs. Johnnie's father would carry their bird in its cage to a park or field. There while Johnnie hunted for insects to feed it, his father traded hints on bird care with other bird-walkers.

The most interesting bird in the shop to Johnnie was the colorful paddy bird. These were small birds, no bigger than sparrows, and people bought them in numbers for only one reason: to set them free. During special festivals families gathered at the temple, opened the cages and released their paddy birds as an act of kindness to please the gods. What a joyful sight it was!

The Bird Shop

John Lim

The Firecracker Shop

John Lir

The Firecracker Shop

Firecrackers were set off to drive away evil spirits and to announce festivals such as the Chinese New Year. A family visit to the Firecracker Shop was Johnnie's favorite shopping.

The shop was small and filled with bright red crackers strung everywhere around it. The crackers were braided very carefully into strands so that one cracker would light the next in a steady stream of bangs. The strands were hung down the front of buildings or from long bamboo poles stuck into the ground. Families and businesses competed with each other to see who could keep the bangs going longest.

Johnnie's father chose their family's crackers with great care, and Johnnie, his brother and sister carried them out like prizes, looking forward to their display. They always hoped their crackers would last the longest in the neighborhood.

The Lantern Maker

Paper lanterns hung everywhere through the streets of Singapore. Usually they were of gentle, subdued shades, but for festivals, they blossomed with color. At the Lantern Shop Johnnie liked to watch them being made. Bamboo strips were shaved and woven into a frame, then sheets of rice paper were cut and pasted over it. Paper lanterns are fragile. They must be carried very carefully and protected from the rain.

During the Moon Festival, Johnnie and the other children received lanterns as gifts. These were the shape of fish, cats, airplanes or flowerpots. When night came, Johnnie's mother carefully positioned a candle inside his lantern and lit it, so he could join the other children who paraded through the neighborhood, showing off their lanterns.

The Kite Shop

After school, if Johnnie had pennies to spend, he headed for the Kite Shop. Like the Lantern Maker, the Kite Maker pasted rice paper (or cellophane) onto a frame of bamboo strips. He then painted colorful designs on it. The fanciest and most expensive kites hung from the ceiling for adult customers who also enjoyed the sport.

Johnnie and the other boys particularly liked kite fighting, although their parents disapproved of it. They put egg yolk and powdered glass into rabbit glue which they applied to the strings of their kites. It made the string cutting sharp. Fighting kites appeared above the rooftops and bobbed up and down, a signal of challenge to a fight. The trick was to cut the string of someone else's kite before he could cut yours.

The Lantern Maker

John Lim

The Kite Shop

John Lim

The Open-air Opera

Of all the wonderful magical things that were sent out into the air around Singapore, nothing matched the music of the Open-air Opera.

As soon as workers arrived to start building the stage for the big event, word spread of the treat in store. Johnnie and his friends rushed down to watch. The stage was built near a temple or marketplace and paid for by a merchant or rich person as a thank offering for good fortune. When the stage was finished, priests from the temple came and blessed it. Then firecrackers were set off to announce the happy occasion.

On the day of the opera, food stalls and peddlers appeared everywhere, selling snacks and drinks. Johnnie and his family came early. Each member carried his own chair and set it up in a favorite spot.

The opera itself was a long, colorful pageant that lasted late into the night. Princes and princesses, warlords and villains, beautifully robed and masked, acted and sang stories of long ago. No matter how hard he tried to stay awake, Johnnie always fell asleep with his head on his mother's lap or against his father's shoulder before it was over.

But the music and excitement seemed to stay in his head and in the air for days afterward.

The Open-air Opera

John Lim

John Lim

John Lim was born in Singapore in 1932 of Chinese parents. Weekends he spent at his grandmother's farm in the village of Bukit Timah outside the city. He described that idyllic experience in his first book AT GRANDMOTHER'S HOUSE, 1977. Weekdays, whenever he wasn't in school, he wandered the streets with his brother, sister and friends, fascinated by the markets, shops, peddlers and artists. This world he now recreates in a new work MERCHANTS OF THE MYSTERIOUS EAST.

He left Singapore after World War II to attend high school in the United States. He graduated from the University of Pennsylvania in 1959, then went to work as an accountant first on Wall Street in New York City, then in England where he became interested in art. His first attempts at sculpture were so successful, he was encouraged to abandon finance for art. Since 1964 he has lived in Toronto, Canada. His sculpture, paintings and limited edition serigraphs are sold at galleries across North America. Posters and postcards, many of them commissioned for art shows, have made him one of the most popular of the new artists.

Singapore

Singapore is both the name of an island in Southeast Asia and the name of its largest city. It is shaped like a diamond, 26 miles long and 14 miles wide. It lies only 85 miles north of the equator. The city of Singapore has long been a port of call for ships, and today it is the fourth largest port in the world. The population of the island is over two million. Seventy-five percent of the people are of Chinese origin, and four major languages are recognized: Malay, Chinese, Tamil and English.

Singapore was a British colony from the time it was founded in 1819 until World War II, when it was occupied by the Japanese. From 1946 on, it passed through various colonial and state forms of government until 1965 when it became an independent republic with a president and a parliament.